By Marty and Sue Bunis

COLLECTOR BOOKS
A Division of Schroeder Publishing Co., Inc.

The current values in this book should be used only as a guide. They are not intended to set prices, which vary from one section of the country to another. Auction prices as well as dealer prices vary greatly and are affected by condition as well as demand. Neither the Authors nor the Publisher assume responsibility for any losses that might be incurred as a result of consulting this guide.

Searching For A Publisher?

We are always looking for knowledgeable people considered to be experts within their fields. If you feel that there is a real need for a book on your collectible subject and have a large comprehensive collection, contact us.

Collector Books
P.O. Box 3009
Paducah, Kentucky 42002-3009

On the Cover:

Raytheon T-100-2, 1956, 3⅜x6⅜x2", $200.00.

Cover design by Beth Summers
Book design by Terri Stalions

Additional copies of this book may be ordered from:

Collector Books
P.O. Box 3009
Paducah, Kentucky 42002-3009

or

Marty and Sue Bunis
RR1, Box 36
Bradford, NH 03221
(603) 938-5051

@$15.95. Add $2.00 for postage and handling.

Printed by IMAGE GRAPHICS, INC., Paducah, Kentucky

Dedication

This book is dedicated to all the radio collectors who have so willingly shared their knowledge and expertise with us for the benefit of radio collectors everywhere. You know who you are, and we thank you.

Acknowledgments

We extend our most sincere thanks to the following people, all of them avid transistor collectors, who provided the transistors radios, photographs, and some of the facts and figures that made this book possible. We couldn't have done this without your help and generosity, and we thank you: Gary Arnold, Bill Burkett, John Clarke, Tim Fuss, Bob Goad, Matt Householder, J. E. Kendall, Steve Lange, Mark Lucas, David Martin, Steven Martin, Dalia Miller, Ken Miller, Larry Mitchell, Kevin Moe, Joe Morinelli, Mark Stein, Jerry Wilson, and Wally Worth.

Special thanks to the following people: Bob Davidson, who was kind enough to invite us into his home twice to photograph his collection; and Bob Evans, who sent dozens of great photographs of his collection.

Sincere thanks also to all those who loaned transistor radio-related paper goods, catalogs, encouragement, and constructive suggestions. We appreciate your help and support.

Introduction

Welcome to the wonderful world of Transistor Radios!

If you have ever wished you could be in on the ground floor of collecting with an item that is still easy to find and affordable but will certainly increase in demand and value in a relatively short time, now is the time to jump on the transistor radio bandwagon. Once viewed by vintage tube radio collectors as second-rate radios barely worth a glance, transistor radios are currently enjoying a rapidly growing popularity and are beginning to take their rightful place in the world of radio collecting. They are plentiful and affordable, and there is still time to build a good collection before prices get too high. In the past year or two, as more and more people discover the fun, excitement, and nostalgia of collecting transistor radios, their popularity has grown at a fantastic rate.

There are several reasons for this recent surge of interest in transistor radio collecting.

1. *Price and availability.* As the older tube sets become harder to find and more expensive, more and more people begin to collect newer, more affordable radios, and transistors fill that bill perfectly. They are still readily available at yard sales and flea markets for relatively low prices, making them a very attractive alternative to older, much more expensive and hard-to-find tube sets. Transistor radio prices will definitely go higher, and now is a great time to buy.

2. *Style and size.* Take a look through the pictures in this book and notice the wonderful styling that is unique to transistor radios. What a change from the old boxy styles of tube table sets and consoles that take up so much space. Transistor radios are small, stylish eyecatchers that are easily displayed in very small spaces.

3. *Nostalgia and historical significance.* First introduced in 1954, transistor radios were an instant success. They were part of the new, exciting space age technology of the fifties and sixties, and many featured futuristic case designs or space age rocket names such as Vanguard, Satellite, Constellation, Galaxy, and Comet. Many of today's new collectors are attracted to transistor radios, simply because they grew up with them and have a nostalgic fondness for collectibles from the space age era.

Explanation of Pricing Section

Because transistor radio collecting is still so new, pricing is much more volatile than that of the older, more established tube sets. One of our main reasons for writing this book is to establish a firm groundwork for transistor radio pricing, one that is sensible and acceptable to a majority of collectors.

The pricing information in this book has been gathered from several sources — classified ads, radio meets, and most notably from a panel of "veteran" transistor radio collectors. Our pricing method is to gather as many prices as possible on each set, discard the high and low ends, and average the remainder to arrive at a fairly current figure. Keep in mind that there is no "suggested retail" price for transistor radios. Pricing is extremely variable and depends almost entirely on the condition of the radio. We have listed pricing that reflects sets in very good condition, using the following guidelines:

1. *Case condition.* One of the great pluses of collecting transistor radios is that they are still so relatively new that they are occasionally found in like-new condition or, better yet, mint in the original box with paperwork. Because most transistor radio collectors want a set to be as close to mint as possible, the condition of the case is of primary concern. As you use this book, keep in mind that the prices written here are for sets in *very good* condition with all parts intact and having no damage to the case. "No damage" is defined as no cracks or hairlines of any kind, no chips, no dents, no heavy scratches (very light wear is acceptable), no deep gouge marks around the coin slot or the seams, and no missing pieces (logos, battery doors, etc). We have not based our pricing on mint sets or those found mint with their original boxes and paperwork — pricing for sets in this pristine, original condition are generally higher than those listed here, usually by up to 100% or more, depending on the scarcity of the set.

2. *Electronically complete.* Transistor radio collectors are generally less concerned with whether a set works or not than they are with the condition and visual appeal of the case, so the chassis of the radio is generally a secondary concern. Although the set may not be working, all the electrical components should be intact and a minimal amount of repair would bring it to operating condition. Before purchasing a transistor radio, be sure to inspect the inside of the case for any unacceptable corrosion from battery leakage, and be sure to remove the batteries from transistor radios on display to prevent any future damage.

Basic Transistor Radio Terms and Descriptive Information

1. *Model Numbers.* We have listed all model numbers in the most logical sequence. Within each company, you will find the numerical model numbers listed first followed by any model numbers beginning with letters. One important point — sometimes, especially with very inexpensive Japanese transistor radios, the original box is the only source of the model number, as some of the cheap sets never had paper labels inside with model information.

If you find a transistor radio at a flea market or yard sale, always be sure to ask if the seller has the box or any original paperwork. It never hurts to ask, and you might end up with more than you bargained for!

2. *Description.* For each set we have included, whenever possible, a general description of the shape of the case (either vertical or horizontal), the material from which the case is made, the approximate date of manufacture, the number of transistors, placement and shape of the dial, knobs and grill, bands, power sources, and any other information important to identifying each set. The following terms are frequently used in the descriptions:

AC – Alternating Current

Airplane Dial – A dial which moves in a 360° circle

AM – Amplitude Modulation

Bat – Battery

Billfold – A transistor radio in the shape of a folding billfold or wallet, usually with the dial, controls, and grill located on the inside of the case

FM – Frequency Modulation

Horizontal – A transistor radio whose case is longer than it is high

Hybrid – The term used for a small group of radios that contain both tubes and transistors. Hybrids radios are popular with both tube and transistor radio collectors, because they represent the transitional period during the mid-1950's when radios were beginning to shift from tubes to transistors.

Leather/Leatherette – The word "leather" is used to describe those cases covered with leather-looking material. In some cases this material may actually be a man-made leatherette.

LW – Long Wave

Measurements – The approximate case measurements included in this book are listed by height first, followed by length, and then depth. They do not include flexible or moving parts, such as swing handles or leather straps.

Slide Rule Dial – A rectangular dial, usually horizontal, which features a thin sliding indicator

SW – Short Wave

Swing Handle – A dual-purpose handle (usually metal) that can be used in an upright position as a handle or in a lower, diagonal position as a stand

Table – A general term used to describe a transistor radio that is a table-top model

Telescoping Antenna – A sectional antenna that folds or telescopes into itself and/or the radio case. Note that some of the telescoping antennas listed here are of the built-in variety that telescope down flush with the case surface and some are of the screw-in kind which only telescope down to the height of their outside shell and must be attached or detached by hand. More often than not, the manually attached type of telescoping antenna is not found with the radio, as they are rather small and easily lost.

Thumbwheel Knob – A thin, round wafer-shaped knob with a serrated outer rim, easily tuned with one finger or thumb

Vertical – A transistor radio whose case is higher than it is wide

Watch Radio – A transistor radio that contains a watch face

Year – We have tried to list the approximate year of manufacture for as many of the sets listed as possible. Please keep in mind that many manufacturers overlapped models from one year to the next, and many popular models were made for a number of years.

3. *Pictures.* There may be a few pictures included here that show transistor radios with some slight case damage, but we have tried, to the best of our ability, to photograph only transistor radios that are as close to original as possible.

Keep in mind that this is a guide only, and it was written to do just that — guide you with identification information and current pricing. Because there are so many variables to consider and because transistor radio prices are escalating rapidly, we make no guarantees that these prices are hard and fast, but we do recommend that you use your judgment when considering the purchase of a transistor radio. If it is in good condition, the price seems fair according to the book, and you like it, buy it!

Feel free to write or call us at any time with anything you feel would be helpful for future transistor radio books. We are always happy to hear from other transistor radio collectors, and we welcome your calls and letters. (If you write, please enclosed a self-addressed, stamped envelope for a reply.) To maintain accuracy, we only list a model number if we have actually seen the radio in person, from a photograph, or vintage company advertisment, so we always welcome information from all sources.

Be sure to check the back of the book for current information on antique radio clubs throughout the country. Radio collecting is growing at a rapid pace, and there are new clubs forming all the time. We recommend joining a club near you. It's a great way to meet others with the same interests.

If you would like to learn more about transistor radios send $1.00 (refundable toward a subscription) for a sample copy of the *Transistor Network,* a monthly newsletter featuring pictures, articles, and classified ads, all exclusively about transistor radios.

Marty & Sue Bunis
RR 1, Box 36
Bradford, NH 03221
(603) 938-5051

Acme

CH-610 "Tops All," horizontal, 1962, six transistors, right front thumbwheel dial, upper right front thumbwheel on/off/volume knob, large perforated grill area, AM, bat **$25.00**

CH-620 "Tops All," horizontal, 1961, six transistors, right front dial with right side thumbwheel tuning, right side thumbwheel on/off/volume knob, perforated grill area, swing handle, AM, bat **$30.00**

Admiral

4P21, horizontal, 1957, black, four transistors, right front round dial knob over large perforated grill area, upper right front thumbwheel on/off/volume knob, chrome swing handle, AM, bat **$35.00**

4P22, horizontal, 1957, red, four transistors, right front round dial knob over large perforated grill area, upper right front thumbwheel on/off/volume knob, chrome swing handle, AM, bat **$40.00**

4P24, horizontal, 1957, tan, four transistors, right front round dial knob over large perforated grill area, upper right front thumbwheel on/off/volume knob, chrome swing handle, AM, bat **$35.00**

4P28, horizontal, 1957, turquoise, four transistors, right front round dial knob over large perforated grill area, upper right front thumbwheel on/off/volume knob, chrome swing handle, AM, bat **$40.00**

7L12, horizontal, 1956, red, six transistors, world's first solar powered radio, runs on battery or on solar power when used with optional "Sun Power Pak," right side dial knob, left side on/off/volume knob, large front grill area with stylized "V," "orbiting electrons" emblem, top pop-up rotating antenna, AM, bat/solar powered.
radio only **$100.00**
radio with Sun Power Pak **$200.00**
radio with Sun Power Pak and leather case **$500.00**

7L14, horizontal, 1956, tan, six transistors, world's first solar powered radio, runs on battery or on solar power when used with optional "Sun Power Pak," right side dial knob, left side on/off/volume knob, large front grill area with stylized "V," "orbiting electrons" emblem, top pop-up rotating antenna, AM, bat/solar powered.
radio only **$100.00**
radio with Sun Power Pak **$200.00**
radio with Sun Power Pak and leather case **$500.00**

7L16, horizontal, 1956, yellow, six transistors, world's first solar powered radio, runs on battery or on solar power when used with optional "Sun Power Pak," right side dial knob, left side on/off/volume knob, large front grill area with stylized "V," "orbiting electrons" emblem, top pop-up rotating antenna, AM, bat/solar powered.
radio only **$100.00**
radio with Sun Power Pak **$200.00**
radio with Sun Power Pak and leather case **$500.00**

7L18, horizontal, 1956, turquoise, six transistors, world's first solar powered radio, runs on battery or on solar power when used with optional "Sun Power Pak," right side dial knob, left side on/off/volume knob, large front grill area with stylized "V," "orbiting electrons" emblem, top pop-up rotating antenna, AM, bat/solar powered.
radio only **$100.00**
radio with Sun Power Pak **$200.00**
radio with Sun Power Pak and leather case **$500.00**

7M12, horizontal, 3⅜x5⅞x1¾", 1958, white and red plastic, seven transistors, right front round dial knob over large metal perforated grill area, upper right front thumbwheel on/off/volume knob, gold swing handle, AM, bat **$50.00**

7M14, horizontal, 3⅜x5⅞x1¾", 1958, white and tan plastic, seven transistors, right front round dial knob over large metal perforated grill area, upper right front thumbwheel on/off/volume knob, gold swing handle, AM, bat $50.00

7M16, horizontal, 3⅜x5⅞x1¾", 1958, white and yellow plastic, seven transistors, right front round dial knob over large metal perforated grill area, upper right front thumbwheel on/off/volume knob, gold swing handle, AM, bat **$50.00**

7M18, horizontal, 3⅜x5⅞x1¾", 1958, white and turquoise plastic, seven transistors, right front round dial knob over large metal perforated grill area, upper right front thumbwheel on/off/volume knob, gold swing handle, AM, bat **$50.00**

221, horizontal, 1958, black, six transistors, upper right front dial knob, upper left front on/off/volume knob, large perforated grill area with lower left stylized "A" logo, handle, AM, bat .. **$35.00**

227, horizontal, 1958, tan, six transistors, upper right front dial knob, upper left front on/off/volume knob, large perforated grill area with lower left stylized "A" logo, handle, AM, bat **$35.00**

228, horizontal, 1958, turquoise, six transistors, upper right front dial knob, upper left front on/off/volume knob, large perforated grill area with lower left stylized "A" logo, handle, AM, bat **$35.00**

521, horizontal, 9x9½x3⅜", 1958, black leatherette, six transistors, upper right front dial knob, upper left front on/off/volume knob, center grill cutouts, dual speakers, rotatable antenna in handle, AM, bat **$30.00**

528, horizontal, 9x9½x3⅜", 1958, blue leatherette, six transistors, upper right front dial knob, upper left front on/off/volume knob, center grill cutouts, dual speakers, rotatable antenna in handle, AM, bat **$35.00**

531, horizontal, 9x9½x3⅜", 1958, charcoal leatherette, eight transistors, upper right front dial knob, two upper left front knobs, center grill cut-outs, dual speakers, rotatable antenna in handle, AM, bat... **$30.00**

537, horizontal, 9x9 1/2x3 3/8", 1958, tan leatherette, eight transistors, upper right front dial knob, two upper left front knobs, center grill cut-outs, dual speakers, rotatable antenna in handle, AM, bat..................... **$30.00**

561 "Super 8," horizontal/table, 1959, plastic, eight transistors, right front dial knob over wrap-around horizontal grill bars, two left front knobs, feet, AM, bat **$25.00**

566 "Super 8," horizontal/table, 1959, plastic, eight transistors, right front dial knob over wrap-around horizontal grill bars, two left front knobs, feet, AM, bat $25.00

581, horizontal, 1959, five transistors, right front dial over large perforated grill area, upper right front thumbwheel knob, AM, bat **$30.00**

582, horizontal, 1959, five transistors, right front dial over large perforated grill area, upper right front thumbwheel knob, AM, bat **$30.00**

692 "Deluxe 5," horizontal, 1960, five transistors, right front dial knob, left front on/off/volume knob, center lattice grill area, crown logo, swing handle, AM, bat..................... **$30.00**

703 "Super 7," horizontal, 3 3/8x5 11/16x1 13/16", 1960, pearl white plastic, seven transistors, right front dial knob, left front on/off/volume knob, center lattice grill area, crown logo, swing handle, AM, bat **$30.00**

708 "Super 7," horizontal, 3 3/8x5 11/16x1 13/16", 1960, Nassau green plastic, seven transistors, right front dial knob, left front on/off/volume knob, center lattice grill area, crown logo, swing handle, AM, bat **$30.00**

711 "Imperial 8," horizontal, 3 3/8x5 11/16x1 13/16", 1959, starlight black/white plastic, eight transistors, right front dial and lower knob, left front knob, center lattice grill area, crown logo, swing handle, AM, bat ...**$30.00**

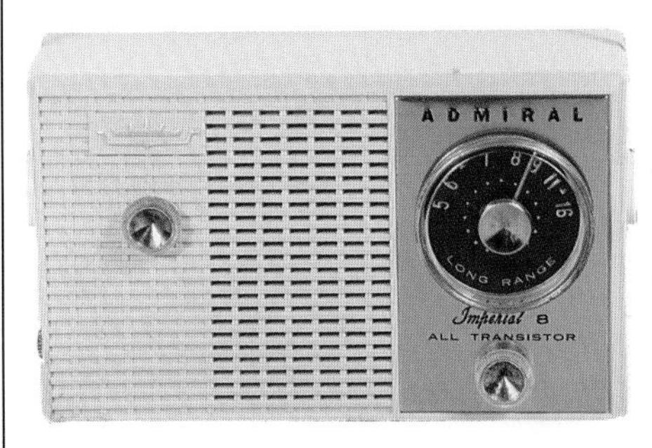

717 "Imperial 8," horizontal, 3 3/8x5 11/16x1 13/16", 1959, Tahiti tan/white plastic, eight transistors, right front dial and lower knob, left front knob, center lattice grill area, crown logo, swing handle, AM, bat $30.00

739, horizontal, 1959, leather, five transistors, right front clear round dial knob over large lattice grill area, left front knob, crown logo, leather handle, AM, bat **$25.00**

742, horizontal, 1959, leather, seven transistors, right front clear round dial knob over large lattice grill area, left front knob, crown logo, leather handle, AM, bat **$25.00**

751, horizontal, 1959, black leather, eight transistors, right front oval window dial, left front knob, large center lattice grill area, crown logo, leather handle, AM, bat **$25.00**

757, horizontal, 1959, tan leather, eight transistors, right front oval window dial, left front knob, large center lattice grill area, crown logo, leather handle, AM, bat **$25.00**

801, horizontal, 1959, eight transistors, right front round dial knob over large perforated grill area, upper right front thumbwheel on/off/volume knob, gold swing handle, AM, bat **$40.00**

808, horizontal, 3⅜x5½x1⅝", 1959, plastic, eight transistors, right front round dial knob over large metal perforated grill area, upper right thumbwheel on/off/volume knob, swing handle, AM, bat.......... **$40.00**

811B "Super 8," horizontal/clock radio, 6¼x11x4", 1959, black back/white front plastic, eight transistors, right front round dial/left front round clock face over wraparound horizontal grill bars, feet, AM, bat **$20.00**

816B "Super 8," horizontal/clock radio, 6¼x11x4", 1959, gold back/white front plastic, eight transistors, right front round dial/left front round clock face with wrap-around horizontal grill bars, feet, AM, bat...... **$20.00**

909 "All World," horizontal, 1960, nine transistors, inner horizontal multi-band slide rule dial, perforated grill area, fold-down front, telescoping antenna, handle, nine bands, bat **$75.00**

PR277, vertical, 4x2½x1¼", plastic, solid state, upper right front window dial with thumbwheel tuning, left side thumbwheel on/off/volume knob, lower perforated plastic grill area, made in Hong Kong, AM, bat $5.00

Y701R, vertical, 4x2½x1¼", plastic, six transistors, upper right front circular window dial, lower vertical grill bars, left side strap, crown logo, AM, bat **$15.00**

Y909 "All World," horizontal, 10x12½x4⅝", 1961, falcon gray leatherette and metal, nine transistors, fold-down front with map and log book, inner horizontal multi-band slide rule dial, telescoping antenna, handle, nine bands, bat **$75.00**

Y2009 "Super 7," horizontal, 3½x5½ x1¾", 1960, plastic, seven transistors, right front dial, left front on/off/volume knob over large lattice grill area, crown logo, AM, bat $25.00

Y2023 "Super 7," horizontal, 1960, white, seven transistors, off-center dial with right and left lattice grill areas, left front knob, crown logo, AM, bat **$25.00**

Y2027 "Super 7," horizontal, 1960, beige, seven transistors, off-center dial with right and left lattice grill areas, left front knob, crown logo, AM, bat **$25.00**

Y2028, "Super 7," horizontal, 1960, green, seven transistors, off-center dial with right and left lattice grill areas, left front knob, crown logo, AM, bat.............................. **$25.00**

Y2061 "Super 7," vertical, 4⅛x2⅝x 1½", 1960, starlight black plastic, seven transistors, upper front off-center window dial with right side thumbwheel tuning, left side thumbwheel on/off/volume knob, lower lattice grill area, crown logo, rear fold-out stand, AM, bat **$20.00**

Y2063 "Super 7," vertical, 4⅛x2⅝x 1½", 1960, pearl white plastic, seven transistors, upper front off-center window dial with right side thumbwheel tuning, left side thumbwheel on/off/volume knob, lower lattice grill area, crown logo, rear fold-out stand, AM, bat $20.00

Y2067 "Super 7," vertical, 4⅛x2⅝x1½", 1960, Sahara beige plastic, seven transistors, upper front off-center window dial with right side thumbwheel tuning, left side thumbwheel on/off/volume knob, lower lattice grill area, crown logo, rear fold-out stand, AM, bat **$20.00**

Y2068 "Super 7," vertical, 4⅛x2⅝x1½", 1960, Nassau green plastic, seven transistors, upper front off-center window dial with right side thumbwheel tuning, left side thumbwheel on/off/volume knob, lower lattice grill area, crown logo, rear fold-out stand, AM, bat **$20.00**

Y2081 "Imperial 7," horizontal, 3½ x5¾ x1⅝", 1961, starlight black plastic, seven transistors, upper right front window dial, lower on/off/volume knob, horizontal front grill bars with crown logo inside large oval, swing handle, made in USA, AM, bat **$20.00**

Y2082 "Imperial 7," horizontal, 3½x5¾ x1⅝", 1961, reef coral plastic, seven transistors, upper right front window dial, lower on/off/volume knob, horizontal front grill bars with crown logo inside large oval, swing handle, made in USA, AM, bat **$20.00**

Y2083 "Imperial 7," horizontal, 3½x 5¾x1⅝", 1961, pearl white plastic, seven transistors, upper right front window dial, lower on/off/volume knob, horizontal front grill bars with crown logo inside large oval, swing handle, made in USA, AM, bat $20.00

Y2091 "Imperial 8," horizontal, 3½x 5⅞x1⅝", 1961, starlight black plastic, eight transistors, upper right front window dial, lower on/off/volume knob, large oval grill area with center crown logo, swing handle, made in USA, AM, bat **$25.00**

Y2093 "Imperial 8," horizontal, 3½x 5⅞x1⅝", 1961, pearl white plastic, eight transistors, upper right front window dial, lower on/off/volume knob, large oval grill area with center crown logo, swing handle, made in USA, AM, bat **$25.00**

Y2098 "Imperial 8," horizontal, 3½x 5⅞x1⅝", 1961, Nassau green plastic, eight transistors, upper right front window dial, lower on/off/volume knob, large oval grill area with center crown logo, swing handle, made in USA, AM, bat **$25.00**

Y2101 "Super 7," vertical, 7½x5⅝x2⅝", 1961, starlight black plastic, seven transistors, upper right front window dial, left on/off/volume knob, large lower grill area with horizontal bars, crown logo, handle, AM, bat .. **$25.00**

Y2102 "Super 7," vertical, 7½x5⅝x2⅝", 1961, reef coral plastic, seven transistors, upper right front window dial, left on/off/volume knob, large lower grill area with horizontal bars, crown logo, handle, AM, bat **$25.00**

Y2108 "Super 7," vertical, 7½x5⅝x 2⅝", 1961, Nassau green plastic, seven transistors, upper right front window dial, left on/off/volume knob, large lower grill area with horizontal bars, crown logo, handle, AM, bat ... **$25.00**

Y2119 "Deluxe 7," horizontal, 5¼x9 x2¾", 1962, dove gray leather, seven transistors, right front dial knob, left horizontal grill bars with on/off/volume knob, crown logo, leather handle, AM, bat..................... **$15.00**

Y2127 "Imperial 8," horizontal, 5¼x 9x2¾", 1959, Tahiti tan leather, eight transistors, right front round dial, left horizontal grill bars with on/off/ volume knob, crown logo, leather handle, AM, bat **$15.00**

Y2137 "Clipper," horizontal, 5¼x9x 3¾", 1961, saddle tan leather, eight transistors, right front semi-circular three band dial and three knobs, left horizontal grill bars with center crown logo, top rotary azimuth scale, leather handle, AM, SW, LW, bat **$20.00**

Y2137C "Clipper," horizontal, 1964, leather, eight transistors, right front semi-circular three band dial and three knobs, left horizontal grill bars with center crown logo, top rotary azimuth scale, leather handle, AM, SW, LW, bat **$20.00**

Y2221, vertical, 3⅜x2½x1", 1962, starlight black plastic, six transistors, upper right front circular window with right side thumbwheel tuning, left side thumbwheel on/off/volume knob, large lower perforated grill area, crown logo, AM, bat **$25.00**

Y2223, vertical, 3⅜x2½x1", 1962, bone white plastic, six transistors, upper right front circular window with right side thumbwheel tuning, left side thumbwheel on/off/volume knob, large lower perforated grill area, crown logo, AM, bat **$25.00**

Y2226, vertical, 3⅜x2½x1", 1962, harvest yellow plastic, six transistors, upper right front circular window with right side thumbwheel tuning, left side thumbwheel on/off/volume knob, large lower perforated grill area, crown logo, AM, bat **$25.00**

Y2229, vertical, 3⅜x2½x1", 1962, jade blue plastic, six transistors, upper right front circular window with right side thumbwheel tuning, left side thumbwheel on/off/volume knob, large lower perforated grill area, crown logo, AM, bat **$25.00**

Y2252, vertical, 7½x5⅝x2⅝", 1962, shell coral plastic, seven transistors, upper right front dial knob, upper left front on/off/volume knob, lower grill area with horizontal bars, handle, AM, bat **$25.00**

Y2253, vertical, 7½x5⅝x2⅝", 1962, bone white plastic, seven transistors, upper right front dial knob, upper left front on/off/volume knob, lower grill area with horizontal bars, handle, AM, bat **$25.00**

Y2256, vertical, 7½x5⅝x2⅝", 1962, harvest yellow plastic, seven transistors, upper right front dial knob, upper left front on/off/volume knob, lower grill area with horizontal bars, handle, AM, bat **$25.00**

Y2271GPN, vertical, 3⅜x2½x1", 1963, starlight black plastic, six transistors, upper right round dial knob over large front textured grill area, AM, bat **$15.00**

Y2272GPN, vertical, 3⅜x2½x1", 1963, ruby red plastic, six transistors, upper right round dial knob over large front textured grill area, AM, bat **$15.00**

Y2273GPN, vertical, 3⅜x2½x1", 1963, bone white plastic, six transistors, upper right round dial knob over large front textured grill area, AM, bat **$15.00**

Y2301GPN, vertical, 3⅜x2½x1", 1963, starlight black plastic, six transistors, upper right front round window dial surrounded by star-shaped trim over large front perforated grill area, right side thumbwheel knob, AM, bat **$30.00**

Y2303GPN, vertical, 3⅜x2½x1", 1963, bone white plastic, six transistors, upper right front round window dial surrounded by star-shaped trim over large front perforated grill area, right side thumbwheel knob, AM, bat **$30.00**

Y2307GPN, vertical, 3⅜x2½x1", 1963, cinnamon brown plastic, six transistors, upper right front round window dial surrounded by star-shaped trim over large front perforated grill area, right side thumbwheel knob, AM, bat **$30.00**

Y2332, horizontal, 6x7½x2½", 1963, ruby red plastic, six transistors, right front round dial knob over large checkered grill area, lower right knob, crown logo, fold-down handle, AM, bat $15.00

Y2351, horizontal, 1963, starlight black leather, eight transistors, upper front horizontal dial, two knobs, large lower grill area with vertical bars, crown logo, leather handle, AM, bat **$15.00**

Y2371, horizontal, 7¼x10⅛x3¾", 1963, chrome and ebony black leatherette, 11 transistors, upper left horizontal two-band slide rule dial, two knobs, large lower perforated grill area, telescoping antenna, handle, AM, FM, bat **$15.00**

Y2401GPN, vertical, 3⅜x2½x1", 1963, six transistors, starlight black plastic, upper right front window dial over large textured grill area, right side thumbwheel knob, AM, bat ... **$15.00**

Y2402GPN, vertical, 3⅜x2½x1", 1963, six transistors, ruby red plastic, upper right front window dial over large textured grill area, right side thumbwheel knob, AM, bat **$20.00**

Y2403GPN, vertical, 3⅜x2½x1", 1963, six transistors, bone white plastic, upper right front window dial over large textured grill area, right side thumbwheel knob, AM, bat ... **$15.00**

Y2411GPN, vertical, 4⅜x2¾x1⅜", 1963, starlight black plastic, eight transistors, large upper right front dial knob, left side thumbwheel on/off/volume knob, lower metal perforated grill area, crown logo, AM, bat **$20.00**

Y2413GP, vertical, 4⅜x2¾x1⅜", 1964, plastic, large upper right front dial knob, left side thumbwheel on/off/volume knob, lower metal perforated grill area, crown logo, AM, bat **$20.00**

Y2413GPN, vertical, 4⅜x2¾x1⅜", 1963, bone white plastic, eight transistors, large upper right front dial knob, left side thumbwheel on/off/volume knob, lower metal perforated grill area, crown logo, AM, bat **$20.00**

Y2421GPN, vertical, 5⅞x3⅝x1½", 1963, black plastic, eight transistors, upper front horizontal slide rule dial with right side thumbwheel tuning, large lower perforated grill area with "starburst" decoration, swing handle, AM, AC/bat **$25.00**

Y2423GPN, vertical, 5⅞x3⅝x1½", 1963, white plastic, eight transistors, upper front horizontal slide rule dial with right side thumbwheel tuning, large lower perforated grill area with "starburst" decoration, swing handle, AM, AC/bat **$25.00**

Y2432, horizontal, 5¼x7¼x2⅜", 1963, ruby red plastic, six transistors, upper right front round dial knob, large lower textured grill area, handle, AM, AC/bat **$10.00**

Y2433, horizontal, 5¼x7¼x2⅜", 1963, bone white plastic, six transistors, upper right front round dial knob, large lower textured grill area, handle, AM, AC/bat **$10.00**

Y2438, horizontal, 5¼x7¼x2⅜", 1963, Nassau green plastic, six transistors, upper right front round dial knob, large lower textured grill area, handle, AM, AC/bat **$10.00**

Y2441, horizontal, 5x7¼x3⅛", 1963, black leather, eight transistors, right front vertical dial and two knobs, left textured grill area with lower left crown logo, leather handle, AM, AC/bat **$10.00**

Y2451, horizontal, 4½x8x1⅞", 1963, jet black plastic, ten transistors, two upper front horizontal slide rule dials – one AM, one FM – right and left thumbwheel knobs, lower textured grill area with lower right switch, telescoping antenna, swing handle, AM, FM, AC/bat **$15.00**

Y2461, horizontal, 5x8x2", 1963, midnight black leather, ten transistors, two upper front horizontal slide rule dials – one AM, one FM – lower textured grill area with lower right band switch, telescoping antenna, leather handle, AM, FM, AC/bat **$15.00**

Y2531GP, vertical, 1965, black front/white back, 10 transistors, upper front horizontal two-band dial, large lower grill area, crown logo, telescoping antenna, swing handle, AM, FM, bat **$15.00**

Y2537GP, vertical, 1965, brown front/white back, 10 transistors, upper front horizontal two-band dial, large lower grill area, crown logo, telescoping antenna, swing handle, AM, FM, bat......................... **$15.00**

Y2539GP, vertical, 1965, blue front/white back, 10 transistors, upper front horizontal two-band dial, large lower grill area, crown logo, telescoping antenna, swing handle, AM, FM, bat **$15.00**

Y2557, horizontal, 5½x7½x3", 1964, tan leather, eight transistors, upper right front dial, large lower grill area with horizontal bars, crown logo, leather handle, AM, bat **$10.00**

Y2577, horizontal, 1965, eight transistors, two upper front horizontal slide rule dials, large lower grill area with lower right AM/SW switch, telescoping antenna, handle, AM, SW, bat **$20.00**

Y2587, horizontal, 1965, eight transistors, upper front horizontal two-band slide rule dial, large lower perforated grill area, telescoping antenna, leather handle, AM, SW, bat ... **$20.00**

YD101GP, vertical, 1965, nine transistors, upper front horizontal two-band slide rule dial, large lower perforated grill area, telescoping antenna, AM, FM, bat **$15.00**

YD107GP, vertical, 1965, nine transistors, upper front horizontal two-band slide rule dial, large lower perforated grill area, telescoping antenna, AM, FM, bat **$15.00**

YD109GP, vertical, 1965, nine transistors, upper front horizontal two-band slide rule dial, large lower perforated grill area, telescoping antenna, AM, FM, bat **$15.00**

YD201GP, horizontal, 2¾x4¼x1", 1965, plastic, upper right front circular window dial with thumbwheel tuning, lower right circular on/off/volume window with thumbwheel knob, left lattice grill area, AM, bat........ **$20.00**

YD242, (top right) horizontal, 5¾x 7¼x2⅜", 1965, plastic, eight transistors, upper right front dial and on/ off/volume knobs, large lower grill area with rectangular slots, fold-down handle, made in Japan, AM, bat $10.00

YH371GP, horizontal, 2⅜x3⅝x1⅛", plastic, eight transistors, upper right front round window dial over large lattice grill area, upper right side thumbwheel dial knob, lower right side thumbwheel on/off/volume knob, made in Japan, AM, bat .. $15.00

Advanco

802 "Super DeLuxe," vertical, 4¼x 2½ x1⅜", plastic, eight transistors, upper right front window dial with thumbwheel tuning, left side thumbwheel on/off/volume knob, lower lattice grill area, made in Hong Kong, AM, bat **$10.00**

MT-608, horizontal, 2¾x4¼x1¼", plastic, six transistors, upper right front square window dial surrounded by concentric squares, horizontal grill bars, top strap, made in Okinawa, AM, bat **$15.00**

Aimor

103, horizontal/travel clock radio, folding leather case, center dial and controls, left perforated grill area, right alarm clock face, AM, FM, bat ... **$20.00**

Air Chief

3-V-80, horizontal, 1963, leather, six transistors, top dial and knob, large front grill area with rectangular cut-outs, leather handle, optional dashboard mounting bracket, AM, bat **$15.00**

4-C-50 "Transiclock," horizontal/clock radio, 1961, seven transistors, lower right front dial knob, upper right lattice grill area, left alarm clock face and on/off/volume knob, swing handle, AM, bat **$20.00**

4-C-55, horizontal, 1964, eight transistors, right front thumbwheel dial, large grill area with horizontal slots, handle, AM, bat **$15.00**

4-C-66, horizontal, 1963, 10 transistors, right front thumbwheel dial, top left thumbwheel on/off/volume knob, large grill area with horizontal slots, handle, AM, bat .. **$15.00**

4-C-69, horizontal, 1963, 14 transistors, upper front horizontal three-band slide rule dial, four knobs, large lower grill area, two telescoping antennas, handle, AM, FM, SW, bat....................... **$20.00**

4-C-96, vertical, 1965, 10 transistors, upper front horizontal two-band slide rule dial, large lower grill area, telescoping antenna, left strap, AM, FM, bat **$10.00**

4-C-97, horizontal, 1965, 10 transistors, right front two-band thumbwheel dial, lower on/off/volume knob, upper AM/FM switch, large left grill area, telescoping antenna, AM, FM, bat **$10.00**

4-C-100, horizontal, 1965, 10 transistors, right front window dial with thumbwheel tuning, large perforated grill area, AM, bat........ **$15.00**

4-C-101, vertical, 1965, 12 transistors, upper front round thumbwheel dial, lower perforated grill area, AM, bat **$15.00**

4-C-102, horizontal, 1965, leather, 12 transistors, right front two-band thumbwheel dial, large perforated grill area, telescoping antenna, handle, AM, FM, bat **$15.00**

Aircastle

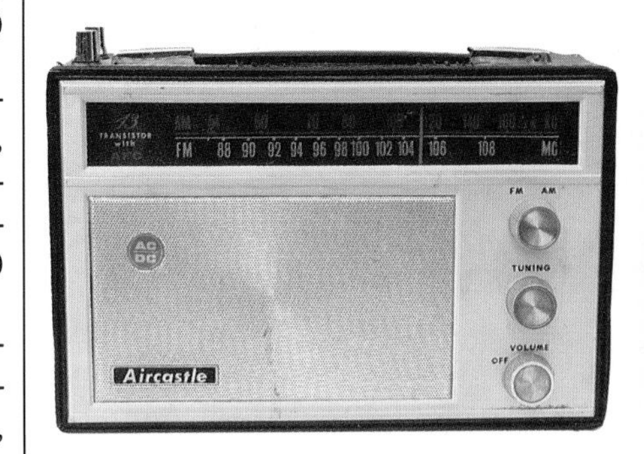

TR1300, horizontal, 5¾x9⅛x2¾", leather, 13 transistors, upper front horizontal two-band slide rule dial, lower metal perforated grill area, two telescoping antennas, rear AC plug, leather handle, made in Japan, AM, FM, AC/bat.......................... $15.00

Airline

BR-1100A, horizontal, right front round dial, lower right thumbwheel on/off/volume knob, center grill area, AM, bat **$175.00**

GEN-1119A, vertical, 1968, plastic, seven transistors, upper right front window dial with right side thumbwheel tuning, left side on/off/volume knob, lower metal perforated grill area, AM, bat $10.00

GEN-1120C, horizontal, 1959, eight transistors, upper right front dial knob, large lower grill area with lower left "M/W" logo, handle, AM, bat **$25.00**

GEN-1131A, horizontal, 1961, six transistors, diagonally divided front, right window dial with right side thumbwheel tuning, right side thumbwheel on/off/volume knob, left perforated grill area, AM, bat **$30.00**

GEN-1136A, vertical, $4\frac{3}{8}$x$2\frac{5}{8}$x$1\frac{1}{4}$", plastic, six transistors, upper right front dial over horizontal grill bars, left circular on/off/volume window with left side thumbwheel knob, made in Taiwan, AM, bat $10.00

GEN-1156B, vertical, $4\frac{1}{2}$x$2\frac{1}{2}$x$1\frac{1}{4}$", plastic, seven transistors, upper right front window dial with right side thumbwheel tuning, upper left on/off/volume window with left side thumbwheel knob, lower metal perforated grill area, AM, bat **$20.00**

GEN-1202A, horizontal, 1962, six transistors, vertically divided front, right oval window dial with right side thumbwheel tuning, right side thumbwheel on/off/volume knob, left perforated grill area with "M/W" logo, AM, bat **$30.00**

GEN-1202B, horizontal, 1963, six transistors, vertically divided front, right oval window dial with right side thumbwheel tuning, right side thumbwheel on/off/volume knob, left perforated grill area with "M/W" logo, AM, bat **$30.00**

GEN-1207A, horizontal, 1961, eight transistors, upper right front dial knob, large lower grill area with lower right on/off/volume knob and lower left "M/W" logo, handle, AM, bat...... **$25.00**

GEN-1208A "Eldorado," horizontal, 1962, six transistors, two right front window dials – one BC, one SW – with right side thumbwheel tuning, lower right on/off/volume window with right side thumbwheel knob, left perforated grill area with upper left logo, telescoping antenna, AM, SW, bat **$30.00**

GEN-1212A, horizontal, 1962, four transistors, right front round dial knob overlaps large perforated grill area, lower right side thumbwheel on/off/volume knob, lower left front "M/W" logo, AM, bat ... **$35.00**

GEN-1213A, vertical, 1963, six transistors, upper front thumbwheel dial, right thumbwheel on/off/volume knob, large lower perforated grill area, AM, bat **$15.00**

GEN-1214A, vertical, 1961, eight transistors, upper left dial knob over perforated front panel, lower right on/off/volume knob, metal swing handle, AM, bat **$30.00**

GEN-1215A, vertical/billfold, 1962, folding billfold style, six transistors, inner right window dial with right side thumbwheel tuning, upper left thumbwheel on/off/volume knob, lower perforated grill area with "M/W" logo, AM, bat **$50.00**

GEN-1218A, vertical, 1962, 10 transistors, upper left front round dial, lower perforated grill area with logo, AM, bat **$20.00**

GEN-1222A, horizontal, 1963, nine transistors, step-back upper front with horizontal two-band slide rule dial, large lower perforated grill area with right knob/left logo, telescoping antenna, handle, AM, FM, bat ... **$20.00**

GEN-1225A, vertical, 1963, six transistors, small upper circular window dial with right side thumbwheel tuning, lower perforated grill area, AM, bat **$20.00**

GEN-1227A, horizontal, 1963, eight transistors, right front window dial and "M/W" logo over large perforated grill area, AM, bat **$15.00**

GEN-1228A, vertical, 1963, eight transistors, upper left front round dial, lower lattice grill with logo, AM, bat **$15.00**

GEN-1229A, horizontal, 1963, 10 transistors, upper right front round dial over large lattice grill area, lower left front knob, swing handle, AM, bat **$15.00**

GEN-1231A, horizontal, 1963, nine transistors, upper front horizontal two-band slide rule dial, large lower perforated grill area with logo, telescoping antenna, handle, AM, FM, bat **$20.00**

GEN-1232A, horizontal, 1964, 10 transistors, upper front horizontal three-band slide rule dial, large lower grill area with logo, telescoping antenna, handle, AM, FM, SW, bat....... **$15.00**

GEN-1240A, vertical, 4¼x2⅝x1¼", 1964, plastic, six transistors, upper right front round window dial with right side thumbwheel tuning, upper left front round on/off/volume window with left side thumbwheel knob, lower metal perforated grill area with center "M/W" logo, made in Japan, AM, bat.................. $20.00

GEN-1245A, horizontal, 1964, leather, eight transistors, upper right front round dial, lower on/off/volume knob, left grill area with horizontal bars, leather handle, AM, bat................ **$10.00**

GEN-1246A, horizontal, 1964, 10 transistors, upper front horizontal two-band slide rule dial, large lower perforated grill area with logo, telescoping antenna, handle, AM, FM, bat **$15.00**

GEN-1247A "Deluxe," horizontal, 1964, 10 transistors, upper front horizontal four-band slide rule dial, large lower perforated grill area, telescoping antenna, handle, AM, FM, 2SW, bat................................ **$20.00**

GEN-1247B "Deluxe," horizontal, 1965, 10 transistors, upper front horizontal four-band slide rule dial, large lower perforated grill area, telescoping antenna, handle, AM, LW, 2SW, bat................................ **$20.00**

GEN-1248A, horizontal, 1964, 12 transistors, upper front horizontal two-band slide rule dial, large lower perforated grill area, handle, AM, FM, bat................................. **$15.00**

GEN-1249A, horizontal, 1964, 14 transistors, upper front horizontal three-band slide rule dial, large lower grill area, two telescoping antennas, handle, AM, FM, SW, bat....... **$15.00**

GEN-1249C, horizontal, 1965, 13 transistors, upper front horizontal three-band slide rule dial, large lower grill area, two telescoping antennas, handle, AM, FM, SW, bat....... **$15.00**

GEN-1253A, vertical, 4¼x2½x1¼", 1965, plastic, seven transistors, upper right front square window dial with right side thumbwheel tuning, upper left on/off/volume window with left side thumbwheel knob, lower metal perforated grill area with center "M/W" logo, AM, bat.................... **$20.00**

GEN-1254A, horizontal, 1965, seven transistors, upper right front window dial with right side thumbwheel tuning, right side thumbwheel on/off/ volume knob, large front perforated grill area with decorative trim and center "M/W" logo, AM, bat ... **$30.00**

GEN-1255A, horizontal, 1965, eight transistors, right front vertical slide rule dial with right side thumbwheel tuning, right side thumbwheel on/ off/volume knob, left grill area with horizontal slots, AM, bat **$15.00**

GEN-1257A, horizontal, 1965, leather, eight transistors, right front vertical slide rule dial and three knobs, left grill area with horizontal bars, leather handle, AM, bat **$10.00**

GEN-1258A, horizontal, 1965, leather, 10 transistors, right front vertical slide rule dial and three knobs, left grill area with horizontal bars, leather handle, AM, bat **$10.00**

GEN-1258B, horizontal, 1965, leather, 10 transistors, right front vertical slide rule dial and three knobs, left grill area with horizontal bars, leather handle, AM, bat **$10.00**

GEN-1259A, vertical, 1964, 10 transistors, upper front horizontal two-band slide rule dial with right side thumbwheel tuning, lower perforated grill area with center "M/W" logo, telescoping antenna, AM, FM, bat **$15.00**

GEN-1260A, horizontal, 1965, leather, 10 transistors, upper front horizontal two-band slide rule dial, large lower grill area with logo, telescoping antenna, leather handle, AM, FM, bat **$10.00**

GEN-1261A, horizontal, 1965, 17 transistors, upper front horizontal four-band slide rule dial, lower left grill area with logo, telescoping antenna, handle, AM, FM, SW, Marine, bat **$20.00**

GEN-1262A, horizontal, 1965, 15 transistors, upper front horizontal six-band slide rule dial, large lower grill area, telescoping antenna, handle, AM, FM, LW, 3SW, bat **$25.00**

GTI-1234A, horizontal, 1963, nine transistors, right front knobs and vertical four-band slide rule dial, left perforated grill area with logo, telescoping antenna, handle, AM, FM, 2SW, bat **$20.00**

GTM-1108A, horizontal, 1958, leather, right front round dial knob, lower thumbwheel on/off/volume knob, left perforated grill area, leather handle, AM, bat **$35.00**

GTM-1109A, horizontal, 1958, seven transistors, right front round dial knob, lower thumbwheel on/off/ volume knob, vertical grill bars, AM, bat **$35.00**

GTM-1200A, horizontal, 1960, nine transistors, upper front horizontal two-band slide rule dial, right knob, large lower grill area with "M/W" logo, telescoping antenna, AM, SW, bat **$20.00**

GTM-1201A, vertical, 1960, seven transistors, lower front round dial over horizontal grill bars, right side on/off/volume knob, swing handle, AM, bat **$25.00**

GTM-1230A, horizontal, 1963, nine transistors, upper front horizontal two-band slide rule dial, large lower grill area with "M/W" logo, telescoping antenna, handle, AM, SW, bat **$15.00**

GTM-1233A, horizontal, 1963, leather, nine transistors, upper front horizontal three-band slide rule dial, top controls, large lower grill area, telescoping antenna, handle, AM, 2SW, bat **$20.00**

Aiwa

AR-102, horizontal, 1964, eight transistors, upper front horizontal four-band slide rule dial, large lower grill area, telescoping antenna, handle, AM, 3SW, bat **$15.00**

AR-111, horizontal, 1964, 11 transistors, upper front horizontal two-band slide rule dial, upper left knob, large lower four-section grill area with off-center knob, telescoping antenna, handle, AM, FM, bat **$20.00**

AR-113, horizontal, 1964, 13 transistors, two upper front horizontal slide rule dials, pushbuttons, top control knobs, lower horizontal grill bars, telescoping antenna, handle, AM, FM, Marine, bat **$20.00**

AR-115, horizontal, 1964, 12 transistors, large right front round two-band dial over horizontal grill bars, telescoping antenna, handle, AM, FM, bat **$20.00**

AR-116, horizontal, 1965, leather, 12 transistors, upper front horizontal two-band slide rule dial, large lower grill area with vertical bars, telescoping antenna, leather handle, AM, FM, bat **$15.00**

AR-122, horizontal, 1965, leather, 12 transistors, two upper front horizontal slide rule dials, large lower grill with vertical bars and left circular grill area, telescoping antenna, leather handle, AM, FM, SW, AC/bat **$20.00**

AR-123, horizontal, 1965, 10 transistors, upper left front horizontal three-band slide rule dial, right side pushbuttons, large lower perforated grill area, telescoping antenna, handle, AM, FM, SW, bat **$15.00**

AR-125, vertical, 1965, 10 transistors, upper front horizontal two-band slide rule dial, large lower grill area with horizontal bars, telescoping antenna, AM, FM, bat **$10.00**

AR-666, vertical, plastic, six transistors, upper right front window dial with right side thumbwheel tuning, upper left front on/off/volume window with left side thumbwheel knob, large metal perforated grill area, AM, bat **$15.00**

AR-670, vertical, 1964, plastic, six transistors, upper right front window dial with right side thumbwheel tuning, left side thumbwheel on/off/volume knob, lower metal perforated grill area with center logo, AM, bat **$15.00**

AR-751, vertical, 1965, seven transistors, upper front horizontal see-through dial, large lower grill area with horizontal bars, AM, bat .. **$20.00**

AR-752, vertical, 4½x3x1½", 1966, plastic, seven transistors, upper front horizontal slide rule dial with thumbwheel tuning, lower lattice grill area, AM, bat **$10.00**

AR-804, horizontal, 1964, eight transistors, upper front horizontal three-band slide rule dial, large lower grill area, telescoping antenna, handle, AM, 2SW, bat **$20.00**

AR-852, horizontal, 1964, eight transistors, right front round dial knob, top thumbwheel knob, left lattice grill with circular speaker area, AM, bat **$25.00**

AR-853, horizontal, 1964, plastic, eight transistors, right front round dial knob, top thumbwheel on/off/volume knob, left grill area with horizontal bars, AM, bat $20.00

AR-854, horizontal, 1964, leather, eight transistors, right front circular two-band dial over vertical grill bars, telescoping antenna, leather handle, AM, SW, bat **$10.00**

Akkord

U60-US "Pinguin," horizontal, 1961, leatherette case with rounded corners, upper front horizontal three-band slide rule dial, pushbuttons, lower grill area with vertical bars, two telescoping antennas, handle, AM, FM, SW, bat **$30.00**

Aladdin

AL65, vertical, 3½x2¼x1", 1962, plastic, six transistors, small upper right front window dial with right side thumbwheel tuning, lower metal perforated grill area with center Aladdin's lamp logo, AM, bat $30.00

AL80, vertical, 1962, eight transistors, upper right front window dial with right side thumbwheel tuning, top left thumbwheel on/off/volume knob, checkered grill area with center Aladdin's lamp logo, AM, bat **$40.00**

Alaron

B-666 "Deluxe HiFi," vertical, 1963, six transistors, upper front window dial with right side thumbwheel tuning, top left thumbwheel on/off/volume knob, lower perforated grill area, AM, bat **$20.00**

DC3280 "Deluxe Eight," horizontal, 1964, nine transistors, upper front horizontal two-band slide rule dial with thumbwheel tuning, lower perforated grill area, telescoping antenna, AM, SW, bat **$25.00**

FAR-113, horizontal, 1964, 12 transistors, upper right front round three-band dial, pushbuttons, large left grill area, left side telescoping antenna, swing handle, AM, FM, SW, bat **$20.00**

TR-709, vertical, 1964, seven transistors, upper right and left side thumbwheel knobs, large front perforated grill area with center logo, AM, bat................................ **$15.00**

TRN-1210, horizontal, 1964, 12 transistors, right front window dial, right side knob, off-center grill area with horizontal slots, AM, bat **$15.00**

TRN-DX, horizontal, 1963, eight transistors, right front airplane dial, two thumbwheel knobs, large left perforated grill area, AM, bat **$15.00**

UR-701, square, 1964, eight transistors, upper right front horizontal dial, two left thumbwheel knobs and one switch, large perforated grill area, AM, bat **$25.00**

Allied

24SC075, vertical, 1965, 10 transistors, upper right front window dial, lower lattice grill area, AM, bat **$5.00**

24SC080, vertical, 10 transistors, upper right front window dial with thumbwheel tuning, lower lattice grill area, AM, bat **$5.00**

TR-1053, vertical, 10 transistors, upper front horizontal slide rule dial with thumbwheel tuning, large lower grill area with vertical bars, AM, bat **$10.00**

Alpha

Q62, vertical, 1962, six transistors, upper right front round dial with thumbwheel tuning, left thumbwheel on/off/volume knob, lower perforated grill area with lower left logo, AM, bat **$40.00**

Ambassador

A-155, horizontal, 1965, 15 transistors, upper front horizontal five-band slide rule dial, large lower grill area, two telescoping antennas, handle, AM, FM, 3SW, bat .. **$20.00**

A-880, horizontal, plastic, eight transistors, right front window dial with thumbwheel tuning, left metal perforated grill area, AM, bat **$25.00**

A-884, horizontal, 1965, eight transistors, right front window dial with thumbwheel tuning, left perforated grill area with upper left logo, AM, bat **$25.00**

A-1064, horizontal, 1965, 10 transistors, right front window dial with thumbwheel tuning, left oval perforated grill area, strap, AM, bat $15.00

AE-10, horizontal, 4⅜x6⅞x2", solid state, upper right front round dial knob, lower on/off/volume knob, left metal perforated grill area with lower left logo, leather handle, made in Hong Kong, AM, AC/bat $10.00

FM-10, vertical, 1965, 10 transistors, upper front horizontal two-band slide rule dial, right side thumbwheel knobs, lower grill area with horizontal slots, telescoping antenna, AM, FM, bat **$20.00**

"Vanguard," horizontal, 5⅛x6⅜x 2½", leather, right side dial knob, left side on/off/volume knob, front grill area with circular cutouts, leather handle, made in USA, AM, bat $20.00

American Supply Co.

61N25-07, vertical, 1961, moonstone plastic, six transistors, diagonally divided front, upper metal panel with right thumb-wheel dial, left side thumbwheel on/off/volume knob, checkered grill area, AM, bat **$25.00**

61N29-07, vertical, 1961, black plastic, six transistors, diagonally divided front, upper metal panel with right thumbwheel dial, left side thumbwheel on/off/volume knob, checkered grill area, AM, bat **$25.00**

Americana

FC60, vertical, 4¼x2½x1¼", 1961, six transistors, upper right front window dial with thumbwheel tuning, top thumbwheel on/off/volume knob, lower perforated grill area, AM, bat $30.00

FM-10, horizontal, 1963, 10 transistors, upper front horizontal two-band slide rule dial, lower perforated grill area with left H/L switch and right AM/FM switch, telescoping antenna, AM, FM, bat **$25.00**

FP62, vertical, 1962, eight transistors, upper front window dial with right side thumbwheel tuning, top left thumbwheel on/off/volume knob, lower perforated grill area, AM, bat **$20.00**

FP64, vertical, 1962, six transistors, upper right front window dial with thumbwheel tuning, lower perforated grill area, made in Japan, AM, bat **$25.00**

FP80, vertical, 1962, plastic, eight transistors, upper right front window dial with right side thumbwheel tuning, top left thumbwheel on/off/volume knob, lower metal perforated grill area, AM, bat $25.00

FP-861, vertical, 1962, eight transistors, upper checkered panel with right thumbwheel window dial, lower wrap-around perforated grill area with lower left logo, AM, bat... **$50.00**

ST-6X "Wayfarer," vertical, 1962, six transistors, upper right front window dial with thumbwheel tuning, upper left thumbwheel on/off/volume knob, lower wrap-around perforated grill area with lower right logo, AM, bat **$40.00**

ST-6Z, vertical, 1962, six transistors, upper right front thumbwheel dial knob, lower perforated grill area with lower left logo, AM, bat **$30.00**

TP-7, horizontal, 1962, leather, seven transistors, upper right front dial knob, left grill with diamond-shaped cut-outs, leather handle, AM, bat **$20.00**

Angel

"Boy's Radio," vertical, 4¼x2⅝x1¼", plastic, two transistors, upper left front round dial knob, upper right front Angel logo, right side thumbwheel on/off/volume knob, lower metal perforated grill area, made in Japan, AM, bat...................... **$40.00**

"Boy's Radio," vertical, 3½x2⅜x1⅛", plastic, two transistors, upper front window dial with top right thumbwheel tuning, top left thumbwheel on/off/volume knob, lower metal perforated grill area with lower left Angel logo, made in Japan, AM, bat **$35.00**

AristoTone

MT-601, vertical, plastic, six transistors, upper right front window dial with thumbwheel tuning, large lower metal perforated grill area, AM, bat $15.00

Artemis

ST-7EL, horizontal, 1961, seven transistors, off-center two-band vertical dial, right side thumbwheel knobs, left front perforated grill area with lower left logo, AM, LW, bat .. **$30.00**

Arvin

60R19, horizontal, 1959, slate gray, four transistors, step-back top, large right front round dial with right thumbwheel on/off/volume knob, left grill area, swing handle, "A" logo, AM, bat **$25.00**

60R23, horizontal, 5x7¼x2¼", 1960, flame plastic, six transistors, step-back top, large right front round dial over vertical grill bars, right thumbwheel on/off/volume knob, swing handle, "A" logo, AM, bat $25.00

60R25-11, compact style, 1½x4⅛x4⅛", plastic, looks like ladies' compact, front window dial with thumb-wheel tuning, rear thumbwheel on/off/volume knob, top metal perforated grill area with floral design and center logo, vinyl wrist strap, made in Hong Kong, AM, bat $25.00

60R25-17, compact style, 1½x4⅛x4⅛", plastic, looks like ladies' compact, front window dial with thumbwheel tuning, rear thumbwheel on/off/volume knob, top metal perforated grill area with floral design and center logo, vinyl wrist strap, made in Hong Kong, AM, bat **$25.00**

60R25-19, compact style, 1½x4⅛ x4⅛", plastic, looks like ladies' compact, front window dial with thumbwheel tuning, rear thumbwheel on/off/volume knob, top metal perforated grill area with floral design and center logo, vinyl wrist strap, made in Hong Kong, AM, bat **$25.00**

60R28, horizontal, 5x7¼x2¼", 1960, sandstone plastic, six transistors, step-back top, large right front round dial over vertical grill bars, right thumbwheel on/off/volume knob, swing handle, "A" logo, AM, bat **$25.00**

60R29, horizontal, 5x7¼x2¼", 1960, gray plastic, six transistors, step-back top, large right front round dial over vertical grill bars, right thumbwheel on/off/volume knob, swing handle, "A" logo, AM, bat **$25.00**

60R33, horizontal, 5x7¼x2⅝", 1959, flame, seven transistors, wedge-shaped case, large right front round dial, right thumbwheel on/off/volume knob, plastic perforated grill area, swing handle, "A" logo, AM, bat **$30.00**

60R35, horizontal, 5x7¼x2⅝", 1959, slate blue, seven transistors, wedge-shaped case, large right front round dial, right thumbwheel on/off/volume knob, plastic perforated grill

area, swing handle, "A" logo, AM, bat $30.00

60R38, horizontal, 5x7¼x2⅝", 1959, taupe, seven transistors, wedge-shaped case, large right front round dial, right thumbwheel on/off/volume knob, plastic perforated grill area, swing handle, "A" logo, AM, bat **$30.00**

60R47, horizontal, 1960, white, seven transistors, upper front off-center thumbwheel dial, right thumbwheel on/off/volume knob, large grill area with horizontal bars and lower left "A" logo, metal handle, AM, bat.... **$30.00**

60R49, horizontal, 1960, black, seven transistors, upper front off-center thumbwheel dial, right thumbwheel on/off/volume knob, large grill area with horizontal bars and lower left "A" logo, metal handle, AM, bat.... **$30.00**

60R58, 1960, tan cowhide, seven transistors, upper right front round dial, left on/off/volume knob, lower checkered grill area, leather handle, AM, bat **$20.00**

60R69, vertical, 1960, black, six transistors, upper right front dial with thumbwheel tuning, left thumbwheel on/off/volume knob, lower perforated grill area, AM, bat... **$20.00**

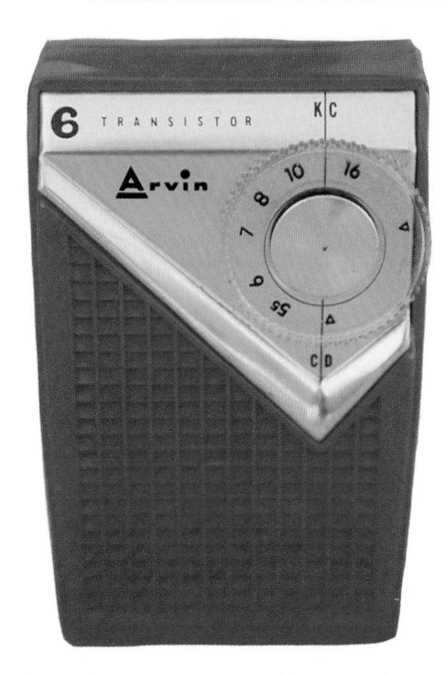

61R13, vertical, 4x2¾x1¼", 1961, red plastic, six transistors, diagonally divided front, upper metal panel with right round dial knob, left side thumbwheel on/off/volume knob, lower checkered grill area, AM, bat $35.00

61R16, vertical, 4x2¾x1¼", 1961, mint green plastic, six transistors, diagonally divided front, upper metal panel with right round dial knob, left side thumbwheel on/off/volume knob, lower checkered grill area, AM, bat **$35.00**

61R19, vertical, 4x2¾x1¼", 1961, black pearl plastic, six transistors, diagonally divided front, upper metal panel with right round dial knob, left side thumbwheel on/off/volume knob, lower checkered grill area, AM, bat **$30.00**

61R23, vertical, 4x2¾x1¼", 1961, sunset plastic, six transistors, diagonally divided front, upper metal panel with right round dial knob, left side thumbwheel on/off/volume knob, lower checkered grill area, AM, bat **$35.00**

61R26, vertical, 4x2¾x1¼", 1961, green plastic, six transistors, diagonally divided front, upper metal panel with right round dial knob, left side thumbwheel on/off/volume knob, lower checkered grill area, AM, bat **$35.00**

61R29, vertical, 4x2¾x1¼", 1961, black plastic, six transistors, diagonally divided front, upper metal panel with right round dial knob, left side thumbwheel on/off/volume knob, lower checkered grill area, AM, bat **$30.00**

61R35, vertical, 4⅛x2½x1⅜", 1961, ice blue plastic, seven transistors, upper right front window dial with thumbwheel tuning, left side thumbwheel on/off/volume knob, lower

perforated grill area, made in USA, AM, bat $25.00

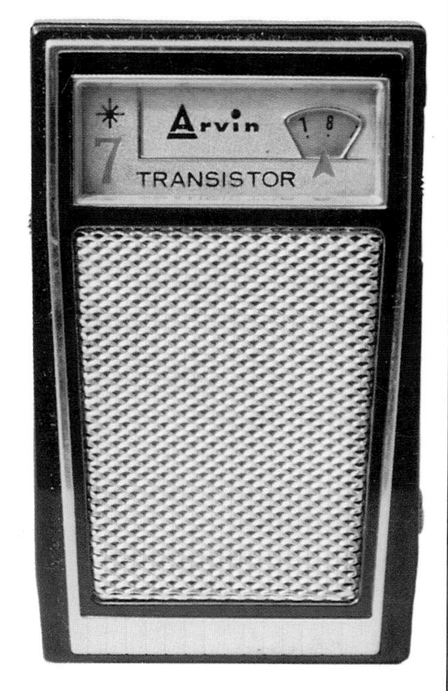

61R39, vertical, 4⅛x2½x1⅜", 1961, jet black plastic, seven transistors, upper right front window dial with thumbwheel tuning, left side thumbwheel on/off/volume knob, lower perforated grill area, made in USA, AM, bat $25.00

61R48, vertical, 1961, leather and metal, seven transistors, upper right front dial with thumbwheel tuning, lower checkered grill area, leather buckle handle, AM, bat **$25.00**

61R58, horizontal, 6x8x3½", 1961, chestnut leather, eight transistors, upper front horizontal slide rule dial, lower metal perforated grill area, leather handle, AM, bat **$15.00**

61R61, vertical, 1963, six transistors, upper right front oversized round dial, lower textured grill area, crown logo, AM, bat **$25.00**

61R64, vertical, 1963, six transistors, upper right front oversized round dial, lower textured grill area, crown logo, AM, bat **$25.00**

61R65, vertical, 1963, six transistors, upper right front oversized round dial, lower textured grill area, crown logo, AM, bat **$25.00**

61R69, vertical, 1963, six transistors, upper right front oversized round dial, lower textured grill area, crown logo, AM, bat **$25.00**

61R79, vertical, 1963, six transistors, upper right front oversized round dial, lower textured grill area, crown logo, AM, bat **$25.00**

61R89 "International II," horizontal, 1961, seven transistors, fold-up front with inner map, inner horizontal three-band slide rule dial, lower checkered grill area, telescoping antenna, handle, AM, SW, LW, bat **$35.00**

61R95, vertical, 1963, six transistors, upper right front oversized round dial, lower textured grill area, crown logo, AM, bat **$25.00**

61R99, vertical, 1963, six transistors, upper right front oversized round dial, lower textured grill area, crown logo, AM, bat **$25.00**

62R09, vertical, 1962, five transistors, upper right front dial with right side thumbwheel tuning, left side thumbwheel on/off/volume knob, vertical grill bars, AM, bat **$15.00**

62R13, vertical, 1962, sunset plastic, six transistors, upper right front dial with right side thumbwheel tuning, left side thumbwheel on/off/volume knob, textured grill area with crown logo, AM, bat **$20.00**

62R16, vertical, 1962, green plastic, six transistors, upper right front dial with right side thumbwheel tuning, left side thumbwheel on/off/volume knob, textured grill area with crown logo, AM, bat **$20.00**

62R19, vertical, 1962, black plastic, six transistors, upper right front dial with right side thumbwheel tuning, left side thumbwheel on/off/volume knob, textured grill area with crown logo, AM, bat **$20.00**

62R23, vertical, 1962, sunset plastic, six transistors, upper right front dial with right side thumbwheel tuning, left side thumbwheel on/off/volume knob, textured grill area with crown logo, AM, bat **$20.00**

62R26, vertical, 1962, green plastic, six transistors, upper right front dial with right side thumbwheel tuning, left side thumbwheel on/off/volume knob, textured grill area with crown logo, AM, bat **$20.00**

62R29, vertical, 1962, black plastic, six transistors, upper right front dial with right side thumbwheel tuning, left side thumbwheel on/off/volume knob, textured grill area with crown logo, AM, bat **$20.00**

62R35, vertical, 1962, blue, seven transistors, upper right front window dial with right side thumbwheel tuning, left side thumbwheel on/off/volume knob, perforated grill area, crown logo, AM, bat **$20.00**

62R39, vertical, 1962, charcoal, seven transistors, upper right front window dial with right side thumbwheel tuning, left side thumbwheel on/off/volume knob, perforated grill area, crown logo, AM, bat **$20.00**

62R48, vertical, 6¼x3½x2", 1962, chestnut leather with metal trim, eight transistors, upper right front round dial with thumbwheel tuning, lower metal perforated grill area, leather buckle handle, AM, bat $25.00

62R49, vertical, 6¼x3½x2", 1962, black leather with metal trim, eight transistors, upper right front round dial with thumbwheel tuning, lower metal perforated grill area, leather buckle handle, AM, bat **$25.00**

62R59, horizontal, 1962, leather, eight transistors, upper front horizontal slide rule dial, lower perforated grill area, leather handle, AM, bat **$15.00**

62R65, vertical, 1962, six transistors, upper right front oversized round dial, lower textured grill area, crown logo, AM, bat **$25.00**

62R69, vertical, 1962, six transistors, upper right front oversized round dial, lower textured grill area, crown logo, AM, bat **$25.00**

62R98, horizontal, 1962, 12 transistors, two upper front round dials – one AM, one FM – lower perforated grill area with right thumbwheel knobs, upper switch, telescoping antenna, handle, AM, FM, bat **$20.00**

63R38, vertical, 1963, leather, seven transistors, upper front horizontal slide rule dial with right side thumbwheel tuning, left side thumbwheel on/off/volume knob, lower perforated grill area, crown logo, AM, bat **$15.00**

63R58, horizontal, 6½x8½x3¾", 1963, brown leather, nine transistors, upper front horizontal slide rule dial, pushbutton dial light, large lower perforated grill area, leather handle, AM, bat **$10.00**

63R88, horizontal, 1963, nine transistors, upper front horizontal three-band slide rule dial, large lower perforated grill area, telescoping antenna, handle, AM, FM, SW, bat **$20.00**

63R98, horizontal, 1963, nine transistors, upper front horizontal four-band slide rule dial, large lower perforated grill area, telescoping antenna, handle, AM, FM, 2SW, bat **$20.00**

64R03, vertical, 1964, six transistors, upper right front window dial with thumbwheel tuning, top left thumbwheel on/off/volume knob, large perforated grill area, AM, bat **$20.00**

64R29, horizontal, 1964, white front/ charcoal back, eight transistors, upper right front window dial with right side thumbwheel tuning, lower right front on/off/volume window with right side thumbwheel knob, large left perforated grill area, AM, bat **$15.00**

64R38, vertical, 4¼x2¾x1½", 1964, walnut leather, eight transistors, upper front horizontal slide rule dial with thumbwheel tuning, left side thumbwheel on/off/volume knob,

large lower metal perforated grill area, detachable braided leather strap, AM, bat $20.00

64R78, horizontal, 1964, 10 transistors, two upper front round dials, AM/FM switch, left thumbwheel on/off/volume knob, lower perforated grill area, telescoping antenna, handle, AM, FM, bat **$20.00**

65R03, vertical, 1965, seven transistors, upper right front window dial with right side thumbwheel tuning, large lower perforated grill area, crown logo, AM, bat **$15.00**

65R03-1, vertical, 1965, seven transistors, upper right front window dial with right side thumbwheel tuning, large lower perforated grill area, crown logo, AM, bat **$15.00**

65R29, vertical, 1965, eight transistors, upper right front window dial with thumbwheel tuning, large lower perforated grill area, crown logo, AM, bat **$15.00**

65R58, horizontal, 1965, brown leather, ten transistors, upper front horizontal slide rule dial, large lower grill area, leather handle, AM, bat . **$15.00**

65R69, vertical, 1965, nine transistors, upper front horizontal two-band slide rule dial, lower grill area with horizontal bars, telescoping antenna, AM, FM, bat **$15.00**

65R79, horizontal, 1965, nine transistors, right front vertical two-band slide rule dial, large left grill area, telescoping antenna, handle, AM, FM, bat **$10.00**

65R98, horizontal, 6½x9x3", 1965, 12 transistors, two upper front round dials, large lower grill area with two right thumbwheel knobs, telescoping antenna, handle, AM, FM, bat **$20.00**

66R29, vertical, 4⅜x2¾x1½", black plastic, eight transistors, upper right front window dial with thumbwheel tuning, left side thumbwheel on/off/volume knob, large metal perforated grill area, made in Hong Kong, AM, bat $15.00

66R39, vertical, 1966, 10 transistors, upper right front window dial, left side thumbwheel knob, large lower perforated grill area, crown logo, AM, bat **$15.00**

66R58, horizontal, 6x7¾x3", 1966, leather, eight transistors, upper front horizontal slide rule dial, large lower plastic grill area, leather handle, AM, bat **$10.00**

66R69, horizontal, 1966, 10 transistors, upper right front horizontal two-band slide rule dial, lower switches and tuning knob, vertical grill bars, telescoping antenna, handle, AM, FM, bat **$10.00**

66R78, horizontal, 1966, 10 transistors, two upper front dials, thumbwheel on/off/volume knob, three switches, perforated grill area, telescoping antenna, handle, crown logo, AM, FM, bat **$20.00**

67R09, vertical, 1967, six transistors, diagonally divided front, upper right front thumbwheel dial, left side thumbwheel on/off/volume knob, lower checkered grill area, AM, bat **$20.00**

67R19, vertical, 1967, six transistors, upper right front dial with right side thumbwheel tuning, left side thumbwheel on/off/volume knob, textured grill area with crown logo, AM, bat **$15.00**

67R29, vertical, 1967, eight transistors, upper right front window dial with thumbwheel tuning, left side thumbwheel on/off/volume knob, large metal perforated grill area, crown logo, AM, bat **$15.00**

67R32, vertical, 1967, 10 transistors, upper right front window dial with thumbwheel tuning, left side thumbwheel on/off/volume knob, large metal perforated grill area, crown logo, AM, bat **$15.00**

68R05, vertical, 1968, six transistors, upper right front window dial with right side thumbwheel tuning, left side thumbwheel on/off/volume knob, lower textured grill area, crown logo, AM, bat **$10.00**

68R38, vertical, 1968, 10 transistors, upper right front window dial, left side thumbwheel knob, large lower perforated grill area, crown logo, AM, bat **$10.00**

68R58, horizontal, 1968, leather, eight transistors, upper front horizontal slide rule dial, large lower grill area, leather handle, crown logo, AM, bat **$10.00**

68R68, horizontal, 1968, leather, 10 transistors, upper front horizontal slide rule dial, large lower perforated grill area, leather handle, crown logo, AM, AC/bat **$10.00**

68R89, horizontal, 1968, leather, eight transistors, upper front horizontal slide rule dial, large lower grill area, handle, crown logo, AM, AC/bat **$10.00**

77R19, vertical, 4½x2¾x1½", 1966, plastic, nine transistors, upper front horizontal two-band slide rule dial, lower perforated grill area, telescoping antenna, AM, FM, bat **$15.00**

77R29, horizontal, 1966, nine transistors, right front vertical two-band slide rule dial, two knobs, large left perforated grill area, telescoping antenna, handle, crown logo, AM, FM, bat **$15.00**

78R09, vertical, 1967, eight transistors, two upper front window dials – one AM, one FM – lower horizontal grill bars, telescoping antenna, AM, FM, bat **$10.00**

78R39, horizontal, 1967, 10 transistors, off-center vertical two-band slide rule dial, three knobs, large left perforated grill area with lower left switch, telescoping antenna, handle, AM, FM, AC/bat **$10.00**

86R19, vertical, 1967, 10 transistors, upper front horizontal three-band slide rule dial, large lower perforated grill area with lower right logo, telescoping antenna, AM, FM, SW, bat **$15.00**

86R29, vertical, 4⅝x2⅞x1⅝", ten transistors, black plastic, upper front horizontal three-band slide rule dial with thumbwheel tuning, right side thumbwheel on/off/volume knob, large lower metal perforated grill area with lower right logo, telescoping antenna, rear SW/AM/FM switch, made in Japan, AM, FM, SW, bat $15.00

87R79, horizontal, 1966, 15 transistors, upper front horizontal four-band slide rule dial, lower perforated grill area, band window, telescoping antenna, handle, AM, FM, SW, Marine, AC/bat **$20.00**

2598, horizontal, 1960, white and charcoal, six transistors, center front round dial over horizontal grill bars, lower left knob, AM, bat **$25.00**

3588, horizontal, 1959, ivory and gray, six transistors, lower left front horizontal slide rule dial, three lower right front knobs, large upper grill area with center logo, feet, AM, bat **$25.00**

7595, horizontal, 5x7⅜x2⅝", 1959, two-tone case, four transistors, large right front dial with thumbwheel knob, left lattice grill area, swing handle, AM, bat **$30.00**

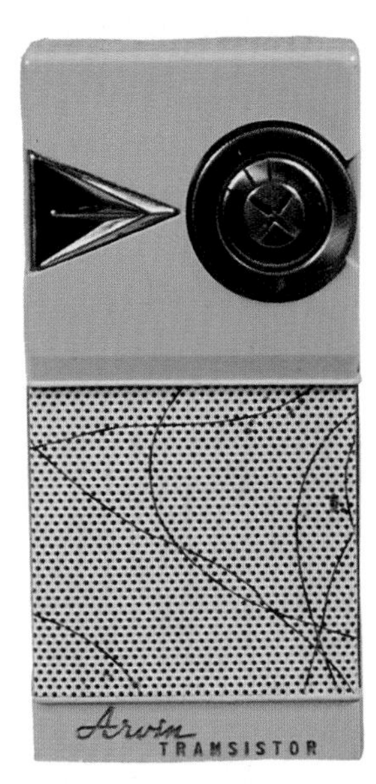

8576, vertical, 1958, available in ebony or turquoise, five transistors, upper right front round dial knob, lower perforated random patterned grill area, rear combination belt clip/stand, AM, bat $135.00

8584, horizontal, 8¾x11⅛x2½", 1959, available in red or blue, five transistors, upper right front thumbwheel dial, upper left thumbwheel on/off/ volume knob, lower grill area with horizontal bars, rotatable antenna in handle, AM, bat $40.00

9562, horizontal, 8⅝x11⅜x3⅞", 1957, available in British tan or alligator, seven transistors, top raised horizontal slide rule dial adjusts for visibility from front or back, two top knobs, large front perforated grill area with brass trim, fold-down handle, made in USA, AM, bat $40.00

9574, horizontal, 8⅜x10¾x3½", 1958, available in white or tan molded case, six transistors, right and left side control knobs, large front lattice grill area with lower right "starburst" emblem, handle, AM, bat **$45.00**

9574-P, horizontal, 8⅜x10¾x3½", 1956, available in white or tan molded case, six transistors, right and left side control knobs, large front lattice grill area with lower right "starburst" emblem, handle, AM, bat **$45.00**

9577, vertical, 1957, available in ebony, pink or white, six transistors, upper front horizontal slide rule dial with right side thumbwheel tuning, left side thumbwheel on/off/volume knob, lower "hourglass" shaped random patterned grill, rear combination belt clip/stand, AM, bat **$100.00**

9594, horizontal, 1960, six transistors, upper front thumbwheel dial, right front thumbwheel on/off/volume knob, left and lower grill area with horizontal bars and "starburst" logo, AM, bat **$35.00**

9595, horizontal, 5x7⅜x2¾", available in white, gray or charcoal wedge-shaped case, seven transistors, large right front round dial over random patterned grill area, right thumbwheel knob, swing handle, AM, bat **$40.00**

9598, horizontal, 1960, seven transistors, upper front horizontal three-band slide rule dial, large lower checkered grill area, telescoping antenna, handle, AM, SW, LW, bat **$20.00**

Astrotone

99-3513L, vertical, 1965, nine transistors, upper right front window dial with thumbwheel tuning, large lower grill area with horizontal bars, AM, bat $10.00

Atkins

61N39-11, vertical, 1964, available in blue or black, seven transistors, upper right front window dial with thumbwheel tuning, left side thumbwheel on/off/volume knob, lower perforated grill area, AM, bat $20.00

61N59-11, horizontal, 1964, black leather, eight transistors, upper front horizontal slide rule dial, large lower grill area, leather handle, AM, bat $10.00

Audition

1069, vertical, plastic, six transistors, upper front V-shaped dial window with top thumbwheel tuning, upper right front thumbwheel on/off/volume knob, lower metal grill area with vertical slots, AM, bat............ $35.00

Automatic

PTR-15B, horizontal, 1958, leather, upper right front dial knob, upper left front on/off/volume knob, large grill area with cut-outs, handle, AM, bat $30.00

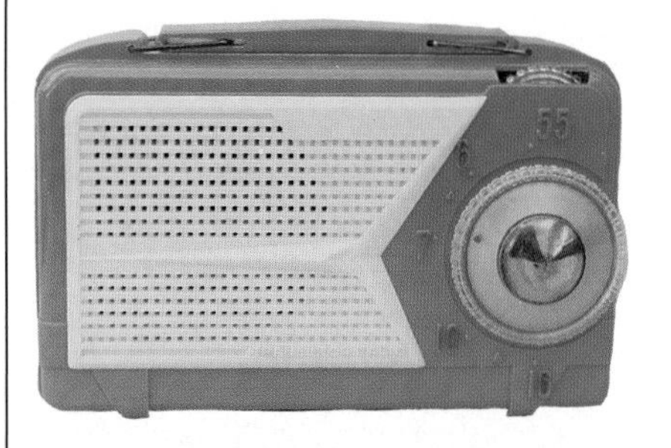

TT 600 "Tom Thumb," horizontal/hybrid, 3¾x5⅞x1⅞", 1955, two-tone plastic, right front round dial knob, top right thumbwheel on/off/volume knob, left checkered grill area, fold-down handle, AM, bat .. $150.00

Barlow

6T-180, vertical, 4¼x2¾x1¼", plastic, six transistors, upper see-through dial with top thumbwheel tuning knob, right side thumbwheel on/off/volume knob, lower metal perforated grill area, AM, bat $45.00

Baylor

6YR-15A, vertical, 1962, six transistors, upper right front thumbwheel dial, upper left front thumbwheel on/off/volume knob, lower round perforated grill area, AM, bat $35.00

Bell Kamra

KTC-62, horizontal/camera radio, plastic, six transistors, right front window dial with right side thumbwheel tuning, right side thumbwheel on/off/volume knob, center round metal perforated grill area, left front camera, made in Japan, AM, bat $100.00

Benida

PR-1161, vertical, 4 x 2½ x 1⅛", plastic, six transistors, upper left front window dial with right side thumbwheel tuning, right side thumbwheel on/off/volume knob, metal perforated grill area with lower right logo, made in Japan, AM, bat $35.00

Blaupunkt

22503 "Lido," horizontal, 1963, nine transistors, upper front horizontal three-band slide rule dial, upper right and left thumbwheel knobs, lower grill area with horizontal bars, telescoping antenna, handle, AM, FM, SW, bat **$25.00**

Bradford

AR-121, vertical, 1965, 10 transistors, step-back top with thumbwheel controls, horizontal two-band slide rule dial, lower grill area with vertical bars, telescoping antenna, AM, FM, bat **$15.00**

AR-857, horizontal, 1964, eight transistors, step-back top with thumbwheel controls, large right front round dial, left grill area, handle, AM, bat **$15.00**

P100, horizontal, 1965, leather, 10 transistors, right front window dial, left grill area with horizontal bars, leather handle, AM, bat **$10.00**

TR-1626, vertical, 4x2½x1¼", 1963, plastic, six transistors, upper right front window dial with right side thumbwheel tuning, left side thumbwheel on/off/volume knob, metal perforated grill area with vertical bar, AM, bat **$30.00**

Browni

702, vertical, 4½x2⅝x1¼", plastic, solid state, upper right front round thumbwheel dial, upper left front round thumbwheel on/off/volume knob, lower grill area with horizontal bars, made in Hong Kong, AM, bat **$10.00**

Buick

981970 "Trans-Portable," horizontal/ auto radio, 3½x7x1½", 1958, metal, designed to be used as a car radio as well as a portable, right front dial over lattice grill area with left Buick logo, left side auto plug, AM .. **$175.00**

Bulova

250/plastic case, vertical, 5x3x1¼", plastic, identical to Regency TR-1 except no earphone jack, upper right front round brass dial knob, upper left front thumbwheel on/off/volume knob, lower perforated grill area, AM, bat $350.00

250 Series/leather case, vertical, 5⅛x3⅜x1⅜", leather, same chassis and dial knob as Regency TR-1, upper right front round brass dial knob, upper left front thumbwheel on/off/volume knob, lower woven cloth grill area, leather strap, AM, bat .. $450.00

260 Series, horizontal, 1957, leather, six transistors, upper right front round dial knob with scalloped edge, right side thumbwheel knob, left checkered grill cut-outs, leather handle, AM, bat **$50.00**

270, horizontal, 1957, leather, right front round dial knob, left plastic three dimensional checkered grill, crown and shield logo, leather handle, AM, bat..................... **$60.00**

278, horizontal, 1958, leather, four transistors, right front round dial knob, right side thumbwheel knob, left plastic three dimensional checkered grill, crown and shield logo, leather handle, AM, bat **$60.00**

290, horizontal, 3x6x1⅝", plastic, right front round tuning knob in crescent-shaped dial area, large checkered grill, right side thumbwheel on/off/volume knob, AM, bat $60.00

290P, horizontal, plastic, right front round tuning knob in crescent-shaped dial area, large checkered grill, right side thumbwheel on/off/volume knob, AM, bat **$60.00**

620, horizontal, 3x6x1½", plastic, right front round thumbwheel dial knob over large crescent shaped checkered grill area with crown and shield logo, swing handle, AM, bat .. $85.00

622, vertical, plastic, six transistors, upper right front window dial with right side thumbwheel tuning, left side thumbwheel on/off/volume knob, lower metal perforated and textured grill area with lower right logo, AM, bat........................ **$30.00**

660 Series, vertical, 5¾x3¾x1¾", 1959, plastic, eight transistors, upper right front window dial with right side

thumbwheel tuning, upper left front on/off/volume window with left side thumbwheel knob, lower metal grill area with raised diamond-shaped cutouts, swing handle, AM, bat... $60.00

672, vertical, 1962, plastic, six transistors, metal front panel with upper right thumbwheel dial and upper left "starburst" decoration with rhinestone, top left switch, lower perforated grill area, AM, bat **$45.00**

685, vertical, 1962, four transistors, upper right front round dial knob, lower grill area with oval shaped cut-outs, swing handle, AM, bat **$35.00**

742 "Super 6," vertical, 1962, six transistors, upper right front thumbwheel dial, lower perforated grill area with lower left logo, AM, bat **$40.00**

782, horizontal, 1962, seven transistors, upper front horizontal two-band slide rule dial with thumbwheel tuning, top left thumbwheel knob, large lower perforated grill area with lower left logo, two switches, telescoping antenna, AM, SW, bat **$35.00**

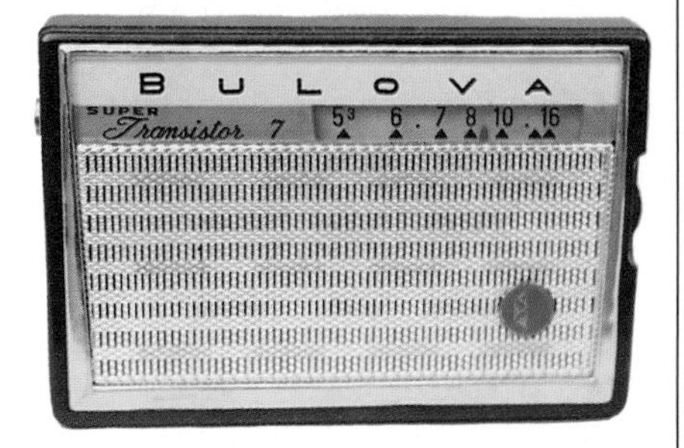

792 "Super Transistor 7," horizontal, 3⅛x4¾x1", 1962, plastic, seven transistors, upper right front horizontal dial with thumbwheel tuning, right side thumbwheel on/off/volume knob, large metal perforated grill area with lower right logo, made in Japan, AM, bat $35.00

840 Series, horizontal/travel watch radio, 3½x5x1½", metal and leatherette folding case, seven transistors, inner metal panel with right round dial, left round watch face, center perforated grill area, AM, bat $40.00

862, horizontal, 1963, nine transistors, upper front horizontal two-band slide rule dial with thumbwheel tuning, left thumbwheel volume knob, top pushbuttons, large lower perforated grill area, telescoping antenna, handle, AM, FM, bat **$30.00**

870, vertical, 3x2x1", 1963, plastic, six transistors, upper right front thumbwheel dial with V-shaped cut-out, lower metal perforated grill area with vertical bars, AM, bat $30.00

872, vertical, 1963, six transistors, upper right front thumbwheel dial with V-shaped cut-out, lower metal perforated grill area with vertical bars, AM, bat......................... **$30.00**

882, horizontal, 1963, seven transistors, upper front horizontal two-band slide rule dial, large lower perforated grill area, telescoping antenna, AM, SW, bat **$30.00**

890 Series, horizontal, 3⅛x4⅞x1", plastic, upper left front horizontal slide rule dial with thumbwheel tuning, right side thumbwheel on/off/volume knob, large perforated grill area, AM, bat **$35.00**

892, horizontal, 1963, seven transistors, upper left front horizontal slide rule dial, two right side thumbwheel knobs, large lower perforated grill area, AM, bat **$35.00**

1042, horizontal, 1965, 12 transistors, upper front horizontal three-band slide rule dial, large lower perforated grill area, telescoping antenna, handle, AM, FM, SW, bat **$20.00**

1130, vertical, 1964, upper front oval window dial with right side thumbwheel tuning, lower perforated grill area, AM, bat **$15.00**

7866, horizontal, 3½x6x1¾", 1962, seven transistors, upper front horizontal two-band slide rule dial with thumbwheel tuning, top left thumbwheel knob, large lower perforated grill area with lower left logo, two switches, telescoping antenna, AM, SW, bat **$30.00**

Calrad

60A183, vertical, 1960, six transistors, upper left front window dial with left side thumbwheel tuning, right side thumbwheel on/off/volume knob, lower perforated grill area with geometric design, AM, bat **$40.00**

Cameo

61N29-03, vertical, 1963, black plastic, six transistors, diagonally divided front, upper metal panel with right round dial knob, left side thumbwheel on/off/volume knob, lower checkered grill area, AM, bat .. **$25.00**

64N06-03, vertical, 1964, six transistors, upper right front round window dial with thumbwheel tuning, top left thumbwheel on/off/volume knob, large perforated grill area, AM, bat **$15.00**

64N09-03, vertical, 1964, six transistors, upper right front window dial with thumbwheel tuning, top left thumbwheel on/off/volume knob, large perforated grill area, AM, bat **$15.00**

Candle

PTR-60S, vertical, 1963, six transistors, upper right front window dial with thumbwheel tuning, lower perforated grill area, "candle" logo, AM, bat **$30.00**

PTR-62B, vertical, plastic, six transistors, upper right front thumbwheel dial, upper left front thumbwheel on/off/volume knob, lower metal perforated grill area, AM, bat **$30.00**

PTR-85C, vertical, 4½x2¾x1⅞", 1963, plastic, eight transistors, step-down top, upper front window dial with upper right thumbwheel tuning, upper left thumbwheel on/off/volume knob, lower metal perforated grill area, made in Japan, AM, bat **$35.00**

PTR-100, horizontal, 3⅛x5¼x1⅜", plastic, 10 transistors, upper right front window dial with top thumbwheel tuning, left side thumbwheel on/

off/volume knob, large metal perforated grill area with lower left logo, made in Japan, AM, bat **$35.00**

Capehart

T6-202 "Incomparable," vertical, 1961, six transistors, upper front panel with crown logo and half-circle window dial, large lower perforated grill area with lower left crown logo, AM, bat **$35.00**

T6-203 "Incomparable," vertical, 1961, six transistors, upper right front window dial with thumbwheel tuning, upper left front on/off/volume window, lower perforated grill area with lower left crown logo, AM, bat **$35.00**

T7-S200 "Incomparable," horizontal, 1961, seven transistors, two right front horizontal dials with thumbwheel tuning, left perforated grill area with lower crown logo, telescoping antenna, AM, SW, bat **$35.00**

T8-201 "Incomparable," vertical, 1961, eight transistors, upper right front round dial with thumbwheel tuning, upper left front thumbwheel knob, lower perforated grill area with lower left logo, swing handle, AM, bat **$45.00**

Casey

010, (top right) vertical, 3¾x2¾x 1½", plastic, solid state, upper right front window dial with top right thumbwheel tuning, top left thumbwheel on/off/volume knob, large lower grill area with horizontal bars, braided strap, made in Hong Kong, AM, bat $5.00

CBS Columbia

TR261, horizontal, 10½x13x6¼" (without stand), wood with brass trim, horizontal two-sided top dial, two top knobs, brass handle, front and rear cloth grill areas, radio snaps off metal stand for use as a table model,

made in USA, AM, bat.
radio without metal stand $50.00
radio with metal stand $75.00

Champion

"Boy's Radio," vertical, 4x2⅜x1⅛", plastic, two transistors, upper right front window dial with right side thumbwheel tuning, left side thumbwheel on/off/volume knob, circular metal perforated grill area with center crest logo, made in Japan, AM, bat $30.00

Champtone

"Boy's Radio," vertical, 4x2½x1⅛", plastic, two transistors, upper front window dial with left side thumbwheel tuning, right side thumbwheel on/off/volume knob, lower round metal perforated grill area, made in Japan, AM, bat $25.00

Channel Master

6459 "Signal Seeker," horizontal, 10 transistors, plastic with metal trim, left front round dial, left top manual thumbwheel tuning knob, right top automatic tuning button, right front perforated grill area, bat $30.00

6474, horizontal, 1965, six transistors, upper right front window dial with thumbwheel tuning, large left metal perforated grill area, AM, bat **$10.00**

6475, vertical, 1965, eight transistors, upper right front round two-band dial with thumbwheel tuning, telescoping antenna, AM, FM, bat **$15.00**

6476 "VHF Monitor," horizontal, 1965, nine transistors, upper front horizontal two-band slide rule dial, large lower grill area with horizontal slots, telescoping antenna, AM, VHF, bat **$20.00**

6477, horizontal, 1965, nine transistors, upper front horizontal two-band slide rule dial, large lower perforated grill area with lower left logo, AM, FM, bat **$15.00**

6479, horizontal, 1965, available in red or black, 14 transistors, upper front horizontal two-band slide rule dial, large lower grill area, telescoping antenna, AM, FM, bat **$15.00**

6500, horizontal, 1960, ivory, six transistors, right front round dial over horizontal grill bars, lower right on/off/volume knob, pull-up handle, feet, AM, bat **$25.00**

6501, horizontal, 3x4¾x1¼", 1959, available in maroon or black, six transistors, upper right front window dial with right side thumbwheel tuning, lower right front on/off/volume window with right side thumbwheel knob, left perforated grill area with logo, AM, bat **$35.00**

6502 (revised), vertical, 4x2½x1¼", 1964, plastic, six transistors, metal front panel with upper right thumbwheel window dial and lower perforated grill area with lower left logo, left side thumbwheel on/off/volume knob, made in Hong Kong, AM, bat $20.00

6503, vertical, 1960, plastic, five transistors, upper right front window dial with right side thumbwheel tuning, lower metal perforated grill area, AM, bat $30.00

6505 "Cordless," 5x10¼x4¾", 1962, caramel and white two-tone plastic, five transistors, upper right front horizontal slide rule dial, large lower checkered grill area with two knobs, fold-down handle, AM, bat $25.00

6506, horizontal, 3x6x1¾", 1960, available in red or black plastic, six transistors, right front dial with thumbwheel tuning, left perforated metal grill area with upper left logo, made in Japan, AM, bat....... **$25.00**

6506A, horizontal, 3x6x1¾", 1960, plastic, six transistors, right front dial with thumbwheel tuning, right side thumbwheel on/off/volume knob, left perforated grill area with upper left logo, made in Japan, AM, bat $25.00

6507, horizontal, 1960, antique ivory and tan two-tone plastic, six transistors, upper right front horizontal two-band slide rule dial, large lower checkered grill area with two knobs and switch, telescoping antenna, fold-down handle, AM, SW, bat **$25.00**

6508 (revised), (bottom left) vertical, 4¼ x2½x1½", 1964, plastic, eight transistors, upper left front round window dial, right side thumbwheel knob, lower metal perforated grill area with center logo, AM, bat $15.00

6509, vertical, 3¾x2¼x1", 1960, available in red or black plastic, six transistors, upper front window dial with thumbwheel tuning, lower metal perforated grill area with center logo, swing handle, AM, bat **$30.00**

6510, horizontal, 5¾x12¾x5", 1960, caramel and white two-tone plastic, six transistors, upper right front horizontal slide rule dial, large lower checkered grill area with two knobs and switch, fold-down handle, AM, bat **$30.00**

6511, horizontal, 6⅛x12¼x5", 1960, Nile green plastic, six transistors, diagonally divided front with upper right see-through slide rule dial and left horizontal grill bars, two knobs, switch, feet, made in Japan, AM, bat $30.00

6512, horizontal, 1960, plastic, eight transistors, right front panel with two-sided rectangular window dial and lower switch, two right side thumbwheel knobs, left perforated grill area with upper left logo, telescoping antenna, AM, SW, bat **$35.00**

6512-2, horizontal, 1960, available in red or black plastic, eight transistors, right front panel with two-sided rectangular window dial and lower switch, two right side thumbwheel knobs, left perforated grill area with upper left logo, telescoping antenna, AM, SW, bat **$35.00**

6514, horizontal, 3½x6¼x1¾", 1960, plastic, eight transistors, right front panel with two-sided rectangular window dial and lower switch, two right side thumbwheel knobs, left perforated grill area with upper left logo, telescoping antenna, AM, Marine, bat **$35.00**

6515 "Super Fringe," horizontal, 4½ x8½x1¾", 1960, available in red or black plastic, eight transistors, right front dial with thumbwheel tuning, right side thumbwheel on/off/volume knob, left metal perforated grill area, AM, bat $25.00

6515-A "Super Fringe," horizontal, 4¼x8½x1¾", plastic, eight transistors, right front dial with thumbwheel tuning, left metal perforated grill area, AM, bat **$25.00**

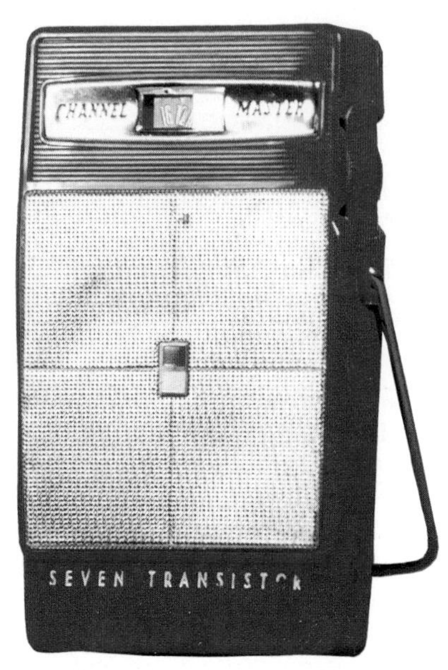

6516, vertical, 1960, available in red or black plastic, seven transistors, upper front window dial with right side thumbwheel tuning, right side thumbwheel on/off/volume knob, lower metal perforated grill area with center logo, swing handle, AM, bat $25.00

6517, horizontal, 1960, black, six transistors, upper front horizontal two-band slide rule dial with thumbwheel tuning, top right on/off/volume knob, large lower metal perforated grill area with BC/LW switch, AM, LW, bat **$30.00**

6518, horizontal, 5½x9½x2¼", 1960, available in red or black plastic, 14 transistors, upper right front horizontal two-band slide rule dial with thumbwheel tuning, thumbwheel on/off/volume knob, large lower metal perforated grill area, two top telescoping antennas, made in Japan, AM, FM, bat **$30.00**

6519, horizontal, 1960, available in red or black plastic, 10 transistors, upper right front horizontal three-band slide rule dial with thumbwheel tuning, right side knob, large lower metal perforated grill area, telescoping antenna, battery test meter, pushbutton dial light, AM, LW, Marine, bat **$30.00**

6520, horizontal, 1960, cream and caramel two-tone plastic, eight transistors, upper right front horizontal slide rule dial, large lower checkered grill area with two knobs and switch, fold-down handle, AM, bat **$25.00**

6521, horizontal/clock radio, 1960, black and chrome, seven transistors, low rectangular radio with center front horizontal slide rule dial and four knobs, top right and left perforated grill areas with center raised alarm clock, telescoping antenna, AM, SW, bat **$25.00**

6522, horizontal, 1963, 10 transistors, right front oval two-band dial, left checkered grill area, four knobs, two telescoping antennas, feet, AM, FM, bat **$25.00**

6523 "Trans-World," horizontal, 1960, available in red or black plastic, 10 transistors, upper right front horizontal three-band slide rule dial, right side knob, large lower metal perforated grill area, telescoping antenna, battery test meter, pushbutton dial light, AM, 2SW, bat **$35.00**

6524, horizontal, 1960, available in red or black plastic, nine transistors, upper front horizontal two-band slide rule dial, right side thumbwheel knob, large lower metal perforated grill area with AM/FM switch, two telescoping antennas, fold-down handle, AM, FM, bat **$25.00**

6527, vertical, 4¼x2½x1¼", 1960, available in red/ivory or black/ivory plastic, six transistors, upper right front window dial with thumbwheel tuning, left side thumbwheel on/off/volume knob, lower lattice grill area, AM, bat $15.00

6528, horizontal, 1960, available in red/ivory or black/ivory plastic, six transistors, upper right front window dial with thumbwheel tuning, large lower lattice grill area, AM, bat **$20.00**

6528A, horizontal, 3x5¼x1½", 1960, plastic, six transistors, upper right front window dial with thumbwheel tuning, large lower lattice grill area with vertical bars, AM, bat **$20.00**

6550 "Swing-Along," portable radio/phono, 4x9x9½", 1962, ivory and caramel plastic with basketweave trim, left front round dial over perforated grill area, five pushbuttons, volume knob, inner top record player with unseen tone arm that plays from underneath the record, inner bottom record storage, flexible plastic handle, AM, bat **$50.00**

6560, horizontal, 1965, eight transistors, right front vertical slide rule dial, two knobs, large left perforated grill area, fold-down handle, AM, bat **$10.00**

6561, horizontal, 1965, six transistors, upper right front round window dial with thumbwheel tuning, thumbwheel on/off/volume knob, lower horizontal grill bars, AM, bat .. **$10.00**

6562, horizontal, 1965, six transistors, right front round dial overlaps vertical grill bars, lower left logo, lower right knob, feet, handle, AM, bat ... **$15.00**

Charmy

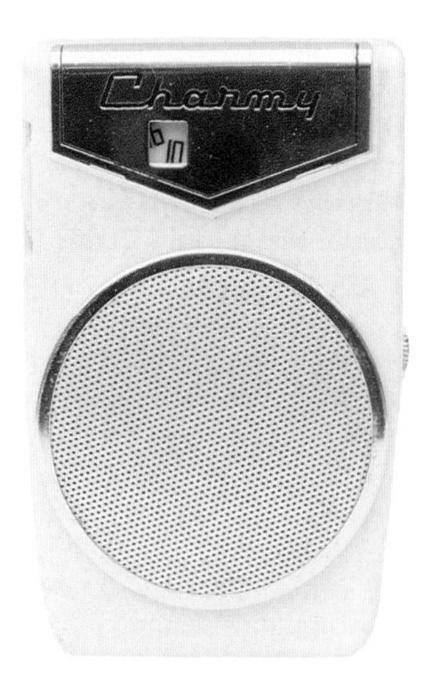

"Boy's Radio," vertical, 3⅞x2½x1⅛", plastic, two transistors, upper left front window dial with left side thumbwheel tuning, right side thumbwheel on/off/volume knob, circular metal perforated grill area, made in Japan, AM, bat $30.00

Clairtone-Braun

T-4, horizontal, 1963, right front window dial, left circular perforated grill area, AM, SW, LW, bat **$20.00**

T22C, horizontal, 1963, nine transistors, upper front horizontal four-band slide rule dial, top pushbuttons and thumbwheel knobs, large lower grill area with horizontal bars, telescoping antenna, handle, AM, FM, 2SW, bat **$25.00**

T-23, horizontal, 1963, eight transistors, upper front horizontal five-band slide rule dial, top pushbuttons and thumbwheel knobs, large lower grill area with horizontal bars, telescoping antenna, handle, AM, 4SW, bat **$25.00**

Claricon

46-070 "Hi-Fidelity," horizontal, 1965, eight transistors, right front round dial and on/off/volume knob, left perforated grill area, leather handle, AM, bat **$15.00**

46-090, horizontal, 1965, 12 transistors, right front round two band dial, and two knobs, left perforated grill area, telescoping antenna in handle, AM, FM, bat **$15.00**

TR605 "VI," vertical, 1964, six transistors, upper right front window dial with thumbwheel tuning, lower grill area with horizontal slots, AM, bat $20.00

Clarion

77A, auto/portable radio, 2x5½x6", plastic/metal, seven transistors, front horizontal slide rule dial, right and left knobs, AM, bat $100.00

Columbia

400B, vertical, 1960, black, four transistors, upper right front window dial with thumbwheel tuning, upper left front on/off/volume window with thumbwheel knob, lower circular grill area with vertical bars, AM, bat **$30.00**

400G, vertical, 1960, gray, four transistors, upper right front window dial with thumbwheel tuning, upper left front on/off/volume window with thumbwheel knob, lower circular grill area with vertical bars, AM, bat **$30.00**

400R, vertical, 1960, red, four transistors, upper right front window dial with thumbwheel tuning, upper left front on/off/volume window with thumbwheel knob, lower circular grill area with vertical bars, AM, bat **$35.00**

600BX, vertical, 1960, black plastic, six transistors, upper right side thumbwheel dial, upper left side thumbwheel on/off/volume knob, lower circular metal perforated grill area, AM, bat **$30.00**

600G, vertical, 1960, green plastic, six transistors, upper right side thumbwheel dial, upper left side thumbwheel on/off/volume knob, lower circular metal perforated grill area, AM, bat **$30.00**

610G "Transistor Convertible," horizontal portable w/speaker box, 1960, gray, six transistors, portable radio unit has right front dial with thumbwheel tuning and left perforated grill area, radio slides into speaker box for use as a table model, AM, bat.
radio without speaker box **$30.00**
radio with speaker box **$60.00**

610R "Transistor Convertible," horizontal portable with speaker box, 1960, red, six transistors, portable radio unit has right front dial with thumbwheel tuning and left perforated grill area, radio slides into speaker box for use as a table model, AM, bat.
radio without speaker box **$35.00**
radio with speaker box **$65.00**

C-605, horizontal, 1962, available in tan or brown leather, five transistors, right side dial knob, left side on/off/volume knob, large front grill area with cut-outs, handle, AM, bat **$15.00**

C-615 "Triumph III," horizontal, 6½ x9x3¼", 1961, plastic leatherette, nine transistors, right front round two-band dial over large metal perforated grill area, four pushbuttons, right side telescoping antenna, handle, AM, FM, bat $30.00

Columbia Records

TR-1000, vertical, 1958, available in saddle tan, rawhide or cordovan leather, six transistors, lift-up top with clasp, inner top right thumbwheel dial and left thumbwheel volume knob, outer front perforated grill area, strap handle, AM, bat ... **$35.00**

Commodore

610A "HiFi," vertical, 3⅞x2⅝x1⅛", plastic, six transistors, upper right front window dial with right side thumbwheel tuning, left side thumbwheel on/off/volume knob, metal perforated grill area, made in Japan, AM, bat...................... **$25.00**

660, vertical, plastic, upper front quarter-round window dial with thumbwheel tuning, left thumbwheel on/off/volume knob, lower metal perforated grill area with center crest logo, AM, bat **$40.00**

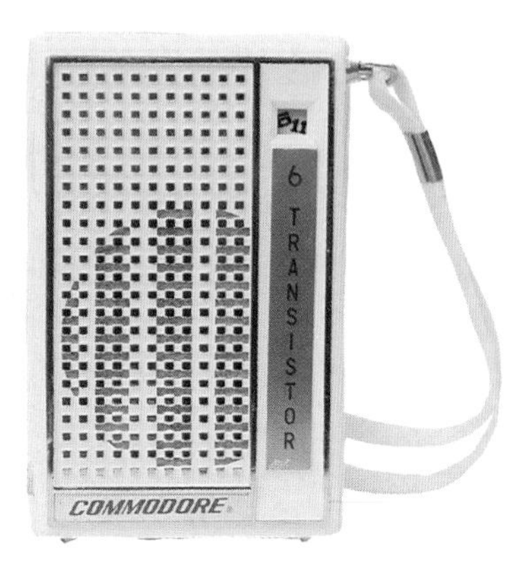

TW-60, vertical, 4x2⅝x1⅛", plastic, six transistors, upper right front window dial with right side thumbwheel tuning, left side thumbwheel on/off/volume knob, large checkered grill area, right side vinyl strap, made in Okinawa, AM, bat $10.00

TW-140, vertical, 4½x2¾x1⅜", plastic, 14 transistors, upper right win-

dow dial with right side thumbwheel tuning, left side thumbwheel on/off/volume knob, lattice grill area, made in Okinawa, AM, bat $10.00

Computron

540 "AA Mark II," horizontal, 2¼x4⅜x1⅛", plastic/metal, top horizontal dial and two knobs, front grill area with horizontal bars, vinyl strap, made in Hong Kong, AM, bat $15.00

Continental

160, vertical, 1959, six transistors, upper left front round dial, upper right front round on/off/volume knob, lower rectangular perforated grill area, center front raised "V", swing handle, AM, bat **$45.00**

MB-7, horizontal, 1961, eight transistors, upper front horizontal three-band slide rule dial with thumbwheel tuning, lower perforated grill area with right knob, telescoping antenna, swing handle, AM, 2SW, bat .. **$25.00**

SW-7, horizontal, 4¼x7x2", 1959, two-tone blue, seven transistors, upper front horizontal three-band slide rule dial with thumbwheel tuning, lower perforated grill area with right knob, swing handle, AM, 2SW, bat .. **$25.00**

TFM-1064, horizontal, 1964, 10 transistors, right front two band dial with thumbwheel tuning, upper left thumbwheel on/off/volume knob, upper right AM/FM switch, large oval perforated grill area, telescoping antenna, AM, FM, bat **$20.00**

TFM-1086, horizontal, 1964, 10 transistors, top horizontal two-band slide rule dial with thumbwheel tuning, three pushbuttons, large front perforated grill area, telescoping antenna, strap, AM, FM, bat **$25.00**

TFM-1087, horizontal, 1964, leather, 10 transistors, right front dial with thumbwheel tuning, upper right AM/FM switch, lower right thumbwheel on/off/volume knob, large left perforated grill area, telescoping antenna, leather handle, AM, FM, AC/bat **$20.00**

TFM-1088, vertical, 1965, nine transistors, upper right front round two-band dial, lower perforated grill area, telescoping antenna, AM, FM, bat **$15.00**

TFM-1090, horizontal, 1964, 10 transistors, right front two-band dial with thumbwheel tuning, large grill area, telescoping antenna, handle, AM, FM, bat.................................. **$15.00**

TFM-1124, horizontal, 8½x12¼x3¾", 1962, black plastic, 11 transistors, upper front horizontal two-band dial with right tuning knob and left tone/volume knobs, large lower grill area, dual speakers, telescoping antenna, AM, FM, bat/optional AC adaptor **$20.00**

TFM-1150-B, horizontal, 5½x9½x2⅝", 1962, 11 transistors, upper front horizontal three-band slide rule dial, large lower perforated grill area, two telescoping antennas, one on either side of the handle, AM, FM, SW, optional AC adaptor/bat **$25.00**

TFM-1155, horizontal, 1964, 11 transistors, upper front horizontal two-band slide rule dial, lower grill area with horizontal bars, telescoping antenna, handle, AM, FM, bat... **$15.00**

TFM-1200, horizontal, 7½x9¼x3", 1965, 12 transistors, upper front horizontal two-band slide rule dial, three knobs, large lower grill area, telescoping antenna, handle, AM, FM, bat................................ **$15.00**

TFM-1365, horizontal, 1964, 13 transistors, upper front horizontal three-band slide rule dial, two left front thumbwheel knobs, two right side knobs, large lower perforated grill area, two telescoping antennas, handle, AM, FM, SW, bat...... **$15.00**

TR-100, vertical, 4⅛x2½x1¼", 1959, two-tone blue, four transistors, upper front window dial with right side thumbwheel tuning, left volume window with left side thumbwheel knob, lower round perforated grill area, swing handle, AM, bat........... **$30.00**

TR-150, horizontal, 3x5½x1½", 1959, available in red or ebony, four transistors, right front dial with right side thumbwheel tuning, lower right volume window with right side thumbwheel knob, left criss-cross grill area, "CMC" logo, AM, bat............. **$40.00**

TR-182, horizontal, 2½x4½x1¼", 1959, available in black or ivory plastic, six transistors, right front window dial with right side thumbwheel tuning, right side thumbwheel on/off/volume knob, large front perforated grill area with upper left logo, AM, bat................................ **$35.00**

TR-200, vertical, 4x2⅜x1¼", 1959, plastic, six transistors, upper right front wedge-shaped dial area with thumbwheel tuning, upper left front on/off/volume window with left side thumbwheel knob, lower perforated grill area, AM, bat................. **$35.00**

TR-208, vertical, 1959, four transistors, upper front window dial with right side thumbwheel tuning, diagonally divided front with left checkered grill area, AM, bat.......... **$35.00**

TR-215, horizontal, 4⅛x5¾x1¼", 1960, available in black or ivory/gray plastic, six transistors, center front vertical dial with thumbwheel tuning, lower right thumbwheel on/off/volume knob, left perforated grill area, AM, bat................................ **$30.00**

TR-300, horizontal, 2¾x4¾x1¼", 1959, eight transistors, right front vertical slide rule dial with right side thumbwheel tuning, lower right on/off/volume window with right side thumbwheel knob, left perforated grill area, AM, bat................. **$25.00**

TR-613, vertical, 1964, six transistors, upper front window dial with right side thumbwheel tuning, top left thumbwheel on/off/volume knob, lower perforated grill area, AM, bat................................ **$20.00**

TR-630, desk set/radio, 1¾x7½x4", 1962, combination pen holder/radio, available in ebony or ivory, seven transistors, front horizontal slide rule dial with thumbwheel tuning, thumbwheel on/off/volume knob, top grill area with diagonal bars, pen holder, AM, bat **$25.00**

TR-632, vertical, 1961, six transistors, upper front window dial with thumbwheel tuning, lower perforated grill area with lower left logo, AM, bat **$35.00**

TR-682, vertical, 4¼x2½x1¼", 1962, plastic, six transistors, center front window dial with top thumbwheel tuning, upper right front thumbwheel on/off/volume knob, lower metal perforated grill area with vertical bars, AM, bat.............. $35.00

TR-716, square, 1965, seven transistors, right side dial and on/off/volume knobs, front perforated grill area with center logo, left side strap, AM, bat **$40.00**

TR-751, vertical, 1961, seven transistors, upper front horizontal two-band slide rule dial with right side thumbwheel tuning, right side thumbwheel on/off/volume knob, lower recessed circular perforated grill area, telescoping antenna, AM, SW, bat **$30.00**

TR-801, horizontal, 3x5x1½", 1961, eight transistors, available in red, gray, or ivory plastic, upper front horizontal slide rule dial with right side thumbwheel tuning, lower right front on/off/volume window with right side thumbwheel knob, large perforated chrome grill area, AM, bat $30.00

TR-814, vertical, 1964, plastic, eight transistors, upper front round window dial with thumbwheel tuning, left side thumbwheel on/off/volume knob, lower metal perforated grill area, AM, bat **$15.00**

TR823 "Globemaster," horizontal, 9x10½x2⅝", 1962, eight transistors, upper front horizontal four-band slide rule dial, large lower grill area with three knobs, telescoping antenna, handle, AM, foreign, 2SW, bat **$25.00**

TR-862, vertical, 1964, eight transistors, upper right front window dial with thumbwheel tuning, lower metal grill area with horizontal slots, AM, bat **$20.00**

TR-875, horizontal, 1964, eight transistors, upper front horizontal three-band wrap-over slide rule dial, large lower perforated grill area, telescoping antenna, AM, 2SW, bat **$25.00**

TR-884, vertical, 4¼x2½x1¼", 1962, eight transistors, upper front window dial with right side thumbwheel tuning, right side thumbwheel on/off/volume knob, large lower perforated grill area, AM, bat **$15.00**

TR-1066, horizontal, 5x9x2¾", 1964, leather, 10 transistors, upper right front dial with thumbwheel tuning, lower right front thumbwheel on/off/volume knob, large center grill area with horizontal slots, leather handle, BC, AC/bat **$15.00**

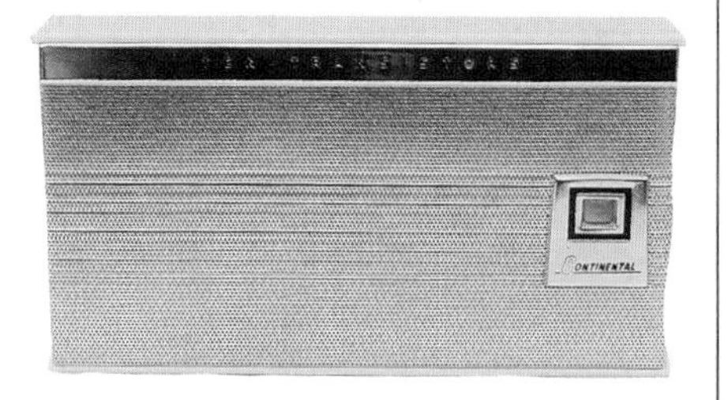

TR-1067, horizontal, 3½x6⅝x1⅝", 1965, ten transistors, right front window dial with right side thumbwheel tuning, lower right side thumbwheel on/off/volume knob, large metal perforated grill area, made in Japan, AM, bat $35.00

TR-1085, vertical, 4½x2¾x1¼", 1964, 10 transistors, upper front window dial with thumbwheel tuning, large lower perforated grill area with center "starburst" logo, AM, bat .. **$20.00**

Coronado

43-9900, vertical, 1960, six transistors, large upper right front window dial with right side thumbwheel tuning, upper left front volume window with thumbwheel knob, lower perforated grill area, AM, bat $30.00

43-9902, horizontal, 1960, six transistors, right front wedge-shaped window dial with right side thumbwheel tuning, right side thumbwheel on/off/volume knob, large perforated grill area, AM, bat **$35.00**

RA44-9914A, vertical, 1963, six transistors, upper right front window dial with right side thumb-wheel tuning, lower perforated grill area, AM, bat **$20.00**

RA44-9915A, horizontal, 1963, seven transistors, upper front thumbwheel dial, left and lower grill area with horizontal bars and lower left logo, swing handle, AM, bat **$30.00**

RA48-9898A, horizontal, 1959, leather, four transistors, upper right front dial knob, upper left front on/off/volume knob, center grill area with cut-outs, leather handle, AM, bat **$30.00**

RA48-9903A "66," horizontal, 1960, six transistors, upper right front round dial, lower on/off/volume knob, left and lower horizontal bars, fold-down handle, AM, bat **$25.00**

RA48-9905A, horizontal, 1960, eight transistors, upper right front round dial, lower on/off/volume knob, left and lower horizontal bars, crown logo, fold-down handle, AM, bat **$25.00**

RA50-9900A, vertical, 1960, six transistors, large upper right front window dial with right side thumbwheel tuning, upper left front volume window with thumbwheel knob, lower perforated grill area, AM, bat **$30.00**

RA50-9902A, horizontal, 1960, six transistors, right front wedge-shaped window dial with right side thumbwheel tuning, right side thumbwheel on/off/volume knob, large perforated grill area, AM, bat **$35.00**

RA60-9899A, vertical, 1962, seven transistors, upper front window dial with right front thumbwheel tuning, left front thumbwheel on/off/volume knob, lower metal perforated wraparound grill area, top and side metal trim, AM, bat **$30.00**

RA60-9917A, horizontal/watch radio, 1963, seven transistors, upper right front horizontal slide rule dial with right side thumbwheel tuning, lower right watch face over large perforated grill area, AM, bat **$75.00**

RA60-9922A, vertical, 1964, six transistors, upper front horizontal slide rule dial with right side thumbwheel tuning, left thumbwheel on/off/volume knob, lower perforated grill area, AM, bat **$15.00**

RA60-9925A, vertical, 1964, six transistors, upper left front window dial with thumbwheel tuning, right and lower checkered grill area, AM, bat **$15.00**

RA60-9940A, horizontal, 1964, 12 transistors, right front round dial, right side knob, left grill area, crown logo, AM, bat **$20.00**

RA60-9941A, horizontal, 1964, eight transistors, right front round dial, lower right knob, large left perforated grill area with center crown logo, handle, AM, bat **$15.00**

RA60-9943A, horizontal, 1964, 12 transistors, upper front horizontal two-band slide rule dial, large lower grill area, telescoping antenna, handle, AM, FM, bat **$15.00**

Coronet

"Boy's Radio," vertical, 4x2½x1¼", plastic, two transistors, upper right front "crown" window dial with right side thumbwheel tuning, left front on/off/volume window with left side thumbwheel knob, metal perforated grill area, telescoping antenna, made in Japan, AM, bat $35.00

Corvair

8P23, vertical, 4¼x2½x1¼", plastic, eight transistors, upper left front round window dial with left side thumb-wheel tuning, right side thumbwheel on/off/volume knob, metal perforated grill area, made in Hong Kong, AM, bat **$25.00**

10PL62, horizontal, 1964, leather, 10 transistors, right front round dial, left grill area, leather handle, AM, bat **$10.00**

10SK63, vertical, 1964, 10 transistors, upper left front round window dial with thumbwheel tuning, lower grill area with horizontal bars, AM, bat **$15.00**

Crest

IV, vertical, 1962, four transistors, upper right front round dial with right side thumbwheel tuning, lower perforated grill area with lower left logo, AM, bat **$35.00**

Crosley

JM-8BG "Musical Memories," book shaped hybrid novelty, 7x4½x2", 1956, leather covered "book," three subminiature tubes and two transistors, outside book spine reads "Musical Memories," inner metal panel with dial, on/off/volume knob and perforated grill area, AM, bat **$150.00**

JM-8BK "Enchantment," book shaped hybrid novelty, 7x4½x2", 1956, leather covered "book," three subminiature tubes and two transistors, outside book spine reads "Enchantment," inner metal panel with dial, on/off/volume knob and perforated grill area, AM, bat **$150.00**

JM-8WE "Treasure Island," (page 59) book shaped hybrid novelty, 7x4½ x2", 1956, leather covered "book," three subminiature tubes and two transistors, outside book spine reads "Treasure Island," inner metal panel with dial, on/off/volume knob and perforated grill area, AM, bat $150.00

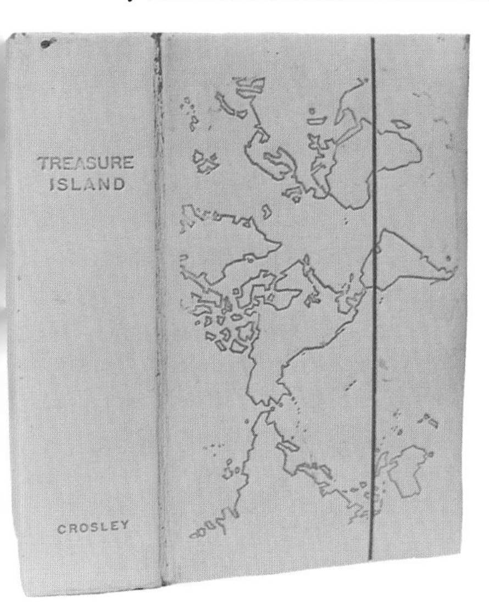

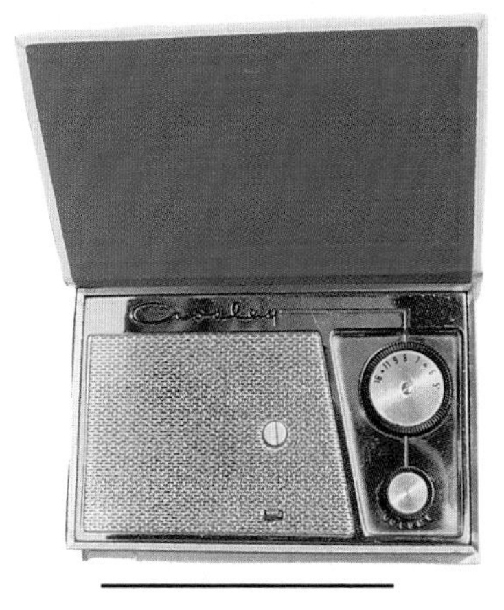

Crown

TR-9, horizontal, 1965, leather, nine transistors, upper right front dial knob, lower right on/off/volume knob, left grill area with horizontal slots, handle, AM, bat **$10.00**

TR-333, vertical, 1959, three transistors, upper right front round dial with right side thumbwheel tuning, left side thumbwheel on/off/volume knob, lower perforated wrap-around grill area, crown logo, AM, bat...... **$45.00**

TR-400, horizontal, 2⅞x4⅞x1½", 1959, available in red, black or blue, four transistors, right front dial with right side thumbwheel tuning, right side thumbwheel on/off/volume knob, left lattice grill area, swing handle, AM, bat **$50.00**

TR-555, vertical, 1960, five transistors, upper front window dial with thumbwheel tuning, lower perforated grill area, crown logo, AM, bat **$40.00**

TR-610, horizontal, 3¾x6¼x2", 1959, available in gray, blue, turquoise, black, or white, six transistors, right front round dial with arrow pointer, lower right side thumbwheel knob, left perforated grill area, swing handle, AM, bat..................... **$25.00**

TR-666, horizontal, 1959, six transistors, right front dial with right side thumbwheel tuning, right side thumbwheel on/off/volume knob, diagonally divided front with left perforated grill area, AM, bat **$45.00**

TR-670, vertical, 1960, seven transistors, small upper front diamond shaped window dial with thumbwheel tuning, lower perforated grill area with lower left crown logo, AM, bat **$35.00**

TR-680, vertical, 3½x2½x1", plastic, six transistors, upper front window dial with thumbwheel tuning, lower metal perforated grill area, AM, bat **$25.00**

TR-750, horizontal, 1961, seven transistors, two right front window dials – one broadcast, one shortwave, right side thumbwheel tuning, lower right front thumbwheel on/off/volume knob, left perforated grill area with lower crown logo, telescoping antenna, AM, SW, bat **$35.00**

TR-777, vertical, 1960, seven transistors, upper front window dial with right side thumbwheel tuning, left side thumbwheel on/off/volume knob, lower perforated grill area with lower left crown logo, AM, bat **$45.00**

TR-800, horizontal, 1960, eight transistors, right front vertical slide rule dial with thumbwheel tuning, lower right front thumbwheel on/off/volume knob, left circular perforated grill area, lower left crown logo, swing handle, AM, bat **$50.00**

TR-820, horizontal, 1959, four transistors, right front round dial, lower right thumbwheel on/off/volume knob, left checkered grill area, swing handle, AM, bat **$40.00**

TR-830, horizontal, 1959, four transistors, right front round dial over horizontal grill bars, lower right thumbwheel on/off/volume knob, AM, bat **$30.00**

TR-875, horizontal, 1960, eight transistors, upper front horizontal two-band slide rule dial, right side knobs, large perforated grill area with lower left crown logo, telescoping antenna, AM, SW, bat **$35.00**

TR-999 "Super 9," vertical, 1961, nine transistors, upper front window dial with right side thumbwheel tuning, right side thumbwheel on/off/volume knob, lower perforated grill area with left crown logo, AM, bat .. **$45.00**

TRF-1000, horizontal, 1965, 13 transistors, upper right front horizontal three-band slide rule dial, large lower grill area, two telescoping antennas, handle, AM, FM, SW, bat **$20.00**

TRF-1600R, horizontal, 1965, nine transistors, upper front horizontal two-band slide rule dial, upper left thumbwheel on/off/volume knob, lower right front AM/FM switch, center grill area with cut-outs, telescoping antenna, AM, FM, bat/rechargeable **$20.00**

TRF-1700, horizontal, 1965, 14 transistors, off-center three-band vertical slide rule dial, right knobs, left perforated grill area, telescoping antenna, handle, AM, FM, SW, bat **$20.00**

TRF-1800, vertical, 1965, nine transistors, upper front horizontal two-band slide rule dial, large lower perforated grill area, telescoping antenna, AM, FM, bat **$15.00**

Dansette

60, horizontal/table, 8½x18x5¼", wood case, large front woven grill area with center round dial and two knobs, metal legs, MW, LW, bat **$30.00**

Delco

989131 "Trans-Portable," auto/portable radio, 2x4¼x6¾", 1958, plastic and metal, designed to be used as a car radio as well as a portable, chrome front with horizontal dial and horizontal grill bars, two knobs, swing handle, AM **$75.00**

Delmonico

6TRS, horizontal, 2½x4½x1¼", plastic, right front window dial with thumbwheel tuning, large metal perforated grill area, AM, bat $25.00

7TA-2, horizontal, 1963, seven transistors, upper right front horizontal two-band slide rule dial, upper left thumbwheel on/off/volume knob, large perforated grill area, telescoping antenna, AM, SW, bat...... **$30.00**

7TH-1, horizontal, 1963, seven transistors, right round two-band dial over large lattice front with left round speaker area, telescoping antenna, handle, AM, SW, bat **$25.00**

7YR707, square, 1965, seven transistors, right side dial and on/off/volume knobs, front circular grill area with center logo, left side chain, AM, bat **$55.00**

8TR8, horizontal, 1965, eight transistors, upper right front window dial with thumbwheel tuning, top thumbwheel on/off/volume knob, front oval perforated grill area, AM, bat **$20.00**

9FM190, horizontal, 1965, nine transistors, upper front horizontal two-band slide rule dial, large lower perforated grill area with lower right logo, telescoping antenna, handle, AM, FM, bat **$15.00**

10TR10, horizontal, 1965, 10 transistors, upper right front window dial with thumbwheel tuning, top thumbwheel on/off/volume knob, front oval perforated grill area, AM, bat **$20.00**

AW6000 "International," 1965, 12 transistors, inner horizontal six-band slide rule dial, large lower grill area, fold-up front, two telescoping antennas, handle, AM, FM, 3SW, LW, bat **$65.00**

TR-7C, horizontal/watch radio, 1963, seven transistors, upper right front horizontal slide rule dial with thumbwheel tuning, lower right front watch face over perforated grill area, AM, bat **$65.00**

TRS-6, horizontal, 1959, six transistors, right front window dial with right side thumbwheel tuning over large grill area with vertical slots, lower right side thumbwheel on/off/volume knob, AM, bat **$30.00**

Deluxe

G-601, horizontal, plastic, "mother-of-pearl" right front with thumbwheel dial knob, top thumbwheel on/off/volume knob, left metal perforated grill area with upper left globe logo, AM, bat **$40.00**

Dewald

K-544, horizontal, 3½x5⅞x2¼", 1957, leather, four transistors, right front round dial knob inside horseshoe-shaped tuning area, upper thumbwheel on/off/volume knob, left grill area with diamond cut-outs and perforations, leather handle, made in USA, AM, bat $125.00

K-544A, horizontal, 3½x5⅞x2¼", 1957, leather, four transistors, right front round dial knob inside horseshoe-shaped tuning area, upper thumbwheel on/off/volume knob, left grill area with diamond cut-outs and perforations, leather handle, AM, bat **$125.00**

K-701A, horizontal, 7x8¾x3½", 1956, plastic, six transistors, lower front round dial knob, right side on/off/volume knob, upper vertical grill bars, pull-up handle, AM, bat **$150.00**

K-701B, (top right) horizontal, 7x8¾x3½", 1956, plastic, six transistors, lower front round dial knob, right side on/off/volume knob, upper vertical grill bars, pull-up handle, made in USA, AM, bat $150.00

K-702B, horizontal, 1957, leather, six transistors, lower front round dial knob, right side on/off/volume knob, upper grill with cut-outs, leather handle, AM, bat **$40.00**

L414, horizontal, 1958, leather, three transistors, right side dial and on/off/volume knobs, large front grill area with vertical cut-outs, leather pull-up handle, AM, bat **$40.00**

L-703, horizontal, 1958, leather, six transistors, right side dial knob, left side on/off/volume knob, large front grill area with cut-outs, leather handle, AM, bat..................... **$35.00**

Ducretet Thomson

TR-854, horizontal, 8x12½x4", leatherette on wood, upper front horizontal slide rule dial, large

lower metal perforated grill area with horizontal bar and center logo, one left side knob, two right side knobs, handle, bat........$25.00

Dumont

900, horizontal, 1963, available in ebony or gray, nine transistors, three decorative front panels made up of concentric rectangles, right front window dial with thumbwheel tuning, swing handle, AM, bat..... **$30.00**

1210, horizontal, 1957, leather, six transistors, upper right front dial knob, lower right on/off/volume knob, left grill area with diamond cut-outs, leather strap, AM, bat..... **$40.00**

Eico

RA-6, horizontal, 1961, leather, six transistors, right side dial knob, large front grill area with cut-outs, leather handle, AM, bat **$15.00**

Electra

"Pee Wee 7," vertical, 2⅝x1⅝x⅞", plastic, seven transistors, upper right front window dial with right side thumbwheel tuning, right side thumbwheel on/off/volume knob, large metal perforated grill area, made in Japan, AM, bat.......... **$40.00**

Electronics Guild

E1000, horizontal/clock radio, 1964, six transistors, right front window dial with upper and lower perforated grill areas, left alarm clock face, AM, AC **$15.00**

Elgin

R-800, vertical, 1964, 10 transistors, upper front horizontal dial with thumbwheel tuning, large lower perforated grill area, AM, bat...... **$15.00**

R-1000B, horizontal, 1964, 10 transistors, upper front horizontal dial, right tuning and on/off/volume knobs, left perforated grill area, AM, bat **$15.00**

R-1000C, horizontal, 1964, 10 transistors, upper front horizontal dial, right tuning and on/off/volume knobs, left perforated grill area, AM, bat **$15.00**

R-1100, horizontal, 1964, 10 transistors, upper front horizontal dial, right tuning and on/off/volume knobs, left perforated grill area, AM, bat **$15.00**

R-1200, horizontal, 1964, 10 transistors, upper front horizontal dial, right tuning, on/off/volume and fine tuning knobs, left perforated grill area, handle, AM, bat **$15.00**

R-1400, vertical, 6⅛x3½x1¾", 1965, plastic, 12 transistors, upper front horizontal two-band slide rule dial, lower perforated grill area, three knobs and AM/FM selector switch, telescoping antenna, made in Japan, AM, FM, bat................ **$15.00**

R-1500, horizontal, 1964, 11 transistors, upper front horizontal two-band slide rule dial, lower perforated grill area, two telescoping antennas, handle, AM, FM, bat **$15.00**

R-1600 "Commander," horizontal, 1965, 11 transistors, upper front horizontal three-band slide rule dial, lower perforated grill area, two telescoping antennas, handle, AM, FM, SW, bat **$20.00**

R-1700, horizontal, 1965, 15 transistors, upper front horizontal five-band slide rule dial, large lower two section perforated grill area, telescoping antenna, handle, AM, FM, 2SW, LW, bat **$25.00**

R-1800, vertical/clock radio, 1964, 10 transistors, off-center vertical slide rule dial, right knobs, left alarm clock face, AM, AC **$10.00**

R-1900, horizontal clock/radio, 1965, 12 transistors, upper right horizontal two-band slide rule dial, left alarm clock face, telescoping antenna, feet, AM, FM, AC **$10.00**

Emerson

31P56, horizontal, 3½x6¾x1½", plastic, eight transistors, right front dial with thumbwheel tuning, upper right thumbwheel on/off/volume knob, large left metal perforated grill area with upper left G-clef logo, right side strap, AM, bat $20.00

31P61, horizontal, 5¼x9x2½", leather, nine transistors, upper front horizontal two-band slide rule dial, lower grill area, right tuning and volume knobs, telescoping antenna, handle, AM, FM, bat $10.00

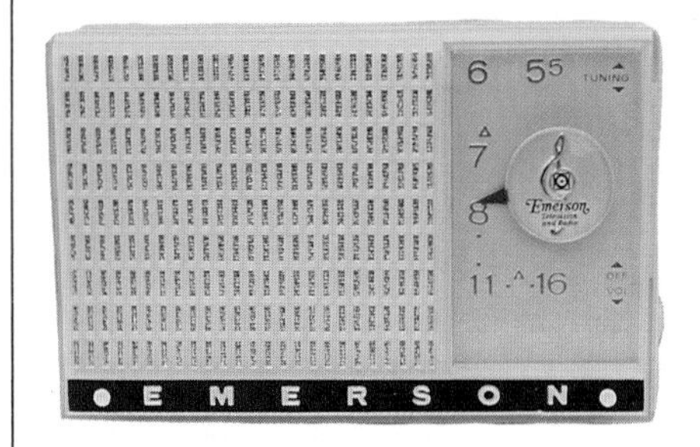

555 "The All-American," horizontal, 3½x6x2", 1959, plastic, four transistors, right front dial with G-clef logo and right side thumbwheel tuning, right side thumbwheel on/off/volume knob, left lattice grill area, "see-through" back, made in USA, AM, bat $55.00

555V, horizontal, 3½x6x2", 1959, plastic, four transistors, right front dial with G-clef logo and right side thumbwheel tuning, right side thumbwheel on/off/volume knob, left lattice grill area, AM, bat... **$55.00**

707, vertical, 4⅛x2½x1¼", 1962, plastic, eight transistors, upper front window dial with top thumbwheel tuning, right side thumbwheel on/off/volume knob, lower metal perforated grill area, made in USA, AM, bat $40.00

808, vertical, 1962, plastic, eight transistors, front window dial with top thumbwheel tuning, right side thumbwheel on/off/volume knob, large front metal perforated grill area, AM, bat **$40.00**

838, horizontal/hybrid, 3¾x6⅜x1½", 1955, plastic, three tubes and two transistors, right front round dial, upper right thumbwheel on/off/volume knob, left lattice grill area with lower left G-clef logo, fold-down handle, AM, bat **$150.00**

842, horizontal, 7x9½x3¾", 1956, leather, six transistors, large center front round dial knob over grill cut-outs, left side on/off/volume knob, leather handle, AM, bat **$35.00**

847, horizontal, 9¾x10x3½", 1957, plastic, six transistors, center front hourglass-shaped checkered grill area with center window dial, right and left front knobs, top "Miracle Wand" antenna in rotatable handle, AM, bat **$40.00**

849, horizontal, 3½x5¾x1½", 1955, plastic, right hourglass-shaped panel with window dial over large front checkered grill area, right side thumbwheel tuning knob, AM, bat $150.00

855, horizontal, 1957, available in red, blue, champagne, cinnamon, or cordovan leather, six transistors, large center front round dial knob over grill cut-outs, leather handle, AM, bat **$40.00**

856, horizontal/hybrid, 3¾x6⅛x1½", plastic, tubes and transistors, right front round dial knob, upper right thumbwheel on/off/volume knob, left checkered grill area, pull-up handle, made in USA, AM, bat.................. **$125.00**

868, horizontal, 9¾x10⅛x3⅜", 1957, plastic, four transistors, upper center front dial, upper left on/off/volume knob, lower grill area with horizontal bars and G-clef logo, top "Miracle Wand" rotatable antenna in handle, AM, bat **$40.00**

869, horizontal, 9¾x10⅛x3⅜", 1957, plastic, four transistors, upper center front dial, upper left on/off/volume knob, lower grill area with horizontal bars and G-clef logo, top "Miracle Wand" rotatable antenna in handle, AM, bat **$40.00**

880, vertical, 1962, eight transistors, front window dial with top thumbwheel tuning, right side thumbwheel on/off/volume knob, lower metal perforated grill area, AM, bat **$40.00**

888 "Atlas," vertical, 6½x4x2", 1960, plastic, eight transistors, upper front round dial knob, left front on/off/volume knob, lower metal perforated random-patterned grill area, swing handle, AM, bat $100.00

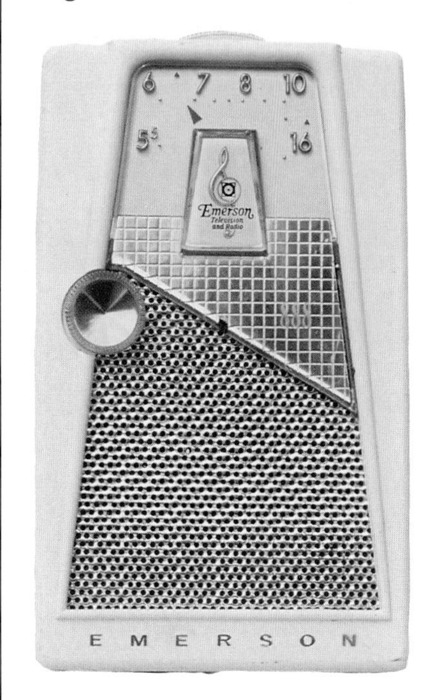

888 "Explorer," vertical, 6½x4x2", 1960, available in white, green, ebony, red, turquoise, or gray plastic, eight transistors, upper front dial with top thumbwheel tuning, diagonally divided lower front with upper checkered plastic panel and lower metal perforated grill area, left front on/off/volume knob, swing handle, made in USA, AM, bat $75.00

888 "Galaxy," vertical/Seattle World's Fair commemorative model, 1963, available in ebony, green, white, red, turquoise, or charcoal plastic, eight transistors, upper front window dial with thumbwheel tuning, right side thumbwheel on/off/volume knob, lower lattice grill area, AM, AC/bat **$70.00**

888 "Pioneer," vertical, 6½x4x2", 1958, plastic, eight transistors, upper front round dial knob, large lower metal perforated grill area with V-shaped top, criss-cross design and lower right G-clef logo, left front on/off/volume knob, swing handle, made in USA, AM, bat $60.00

888 "Satellite," vertical, 1958, available in oyster white or autumn brown leather, eight transistors, upper front round dial knob, left front on/off/volume knob, lower grill area with random-patterned circular cut-outs of varying sizes, leather handle, AM, bat **$150.00**

888 "Titan," vertical, 6x3¾x1¾", 1963, available in ebony, green, white, red, turquoise, or charcoal plastic, eight transistors, off-center oval dial area with center window dial, top thumbwheel tuning knob, right side thumbwheel on/off/volume knob, large front perforated grill area, AM, AC/bat......... **$65.00**

888 "Transtimer," horizontal/clock radio, 7x8⅝x3⅜", 1958, saddle tan leather, eight transistors, front cover unsnaps and folds back to form easel, inner lower left round dial, lower right round metal perforated grill, upper round clock face, leather handle, made in USA, AM, bat.................. **$125.00**

888 "Transtimer II," horizontal/clock radio, 7x8⅝x3⅜", 1959, available in black or tan leather, eight transistors, front cover unsnaps and folds back to form easel, inner lower left round dial, lower right round metal perforated grill, upper round clock face, leather handle, AM, bat $125.00

888 "Vanguard," vertical, 6½x4x2", 1958, available in white, green, ebony, red, or turquoise plastic, eight transistors, futuristic front case design with upper round dial knob, left front on/off/volume knob, lower grill area with random-pattern cut-outs, rocket logo, made in USA, AM, bat **$75.00**

899 "Mercury," vertical, 1964, available in ebony, green, white, red, turquoise, or charcoal, upper front window dial with top thumbwheel tuning, right side thumbwheel on/off/ volume knob, lower lattice grill area, G-clef logo, AM, bat $45.00

911 "Eldorado," horizontal, 4¼x6¾ x2", 1960, available in ebony, charcoal gray, coral, ivory, gold, or turquoise plastic, nine transistors, right front window dial with right side thumbwheel tuning over large checkered grill area, right side on/ off/volume knob, swing handle, made in USA, AM, bat......... **$40.00**

977 "Falcon," horizontal, 1961, seven transistors, right front dial over horizontal front bars, right side thumbwheel on/off/volume knob, AM, bat **$25.00**

988 "Rambler," vertical, 6½x4x2", 1960, available in white, green, ebony, red, or turquoise plastic, eight transistors, upper front dial with top thumbwheel tuning, lower horizontal grill bars, left front on/off/ volume knob, swing handle, made in USA, AM, bat $75.00

991, horizontal, 5¾x8½x3¼", 1963, available in brown, ebony, or charcoal gray leather, nine transistors, upper left front horizontal slide rule dial, lower lattice grill area, right side knob, leather handle, AM, bat **$15.00**

999 "Champion," 1959, plastic, four transistors, upper front round dial knob over checkered grill area, upper right thumbwheel on/off/volume knob, AM, bat **$150.00**

Encore

"Boy's Radio," vertical, 4x2½x1", plastic, two transistors, upper left window dial with thumbwheel tuning, lower round metal perforated grill area, AM, bat **$35.00**

Essex

TR-6K, vertical, six transistors, upper right front window dial with thumbwheel tuning, lower perforated grill area, made in Taiwan, AM, bat **$15.00**

TR-10P "Super DeLuxe," horizontal, 1964, 10 transistors, right front window dial with right side thumbwheel tuning, right side thumbwheel on/off/volume knob, large left perforated grill area, AM, bat **$20.00**

Ever-Play

100, horizontal, 4⅛x8¾x2", 1963, plastic, nine transistors, rechargeable, upper front horizontal two-band slide rule dial, recessed right front with thumbwheel knobs, upper left battery/tuning indicator, large metal perforated grill area, left side telescoping antenna, right side pull-out handle, AM, FM, AC/bat $35.00

1836A "Rechargeable," vertical, 1963, six transistors, upper left front round window dial with right side thumbwheel tuning, right side on/off/volume knob, metal perforated grill area, lower right side recharger plug, AM, bat **$35.00**

2836A, vertical, 1963, eight transistors, rechargeable, upper front horizontal slide rule dial with right side thumbwheel tuning, perforated grill area with logo, AM, AC/bat ... **$35.00**

PR-1266 "Rechargeable," vertical, 4⅜x2¾x1⅜", plastic, six transistors, upper left front round window dial with right side thumbwheel tuning, right side on/off/volume knob, metal perforated grill area, lower right side recharger plug-in, made in Japan, AM, bat $35.00

Excel

6T-2 "Aristocrat," horizontal, 1959, six transistors, right front see-through dial with right side thumbwheel tuning, lower right side thumbwheel on/off/volume knob, left checkered grill area, AM, bat **$35.00**

Executive

CPR-7 "The Executive Desk Radio," horizontal desk set/radio, 1964, seven transistors, top left thumb-wheel knobs and perforated grill area, two pen holders, AM, bat **$25.00**

Faircrest

1094, vertical, 1965, 10 transistors, upper right front window dial, lower lattice grill area, AM, bat **$10.00**

1670, vertical, 1965, six transistors, upper right front window dial, lower grill area with horizontal bars, AM, bat **$15.00**

2091, horizontal, 1965, 10 transistors, upper front horizontal two-band slide rule dial, large lower perforated grill area with lower right AM/FM switch, telescoping antenna, AM, FM, bat **$15.00**

Falcon

6THK, vertical, 4x2½x1¼", 1964, plastic, six transistors, upper right front diamond shaped window dial with right side thumbwheel tuning, left side thumbwheel on/off/volume knob, lower metal perforated grill area, made in Hong Kong, AM, bat **$20.00**

8THK "DeLuxe," vertical, eight transistors, upper right front diamond shaped window dial with right side thumbwheel tuning, left side thumbwheel on/off/volume knob, lower metal perforated grill area with center diamond shaped logo, AM, bat **$25.00**

Firestone

4-C-33, horizontal, 1958, leather, six transistors, upper right front dial knob, upper left front on/off/volume knob, center grill cut-outs, leather handle, AM, bat **$25.00**

4-C-34, horizontal, 1957, British tan leather, seven transistors, horizontal slide rule dial, right and left knobs, large lattice grill area, handle, AM, bat **$40.00**

4-C-36, horizontal, 1959, six transistors, top right dial, top left on/off/volume knob, large front grill area with horizontal bars and lower right logo, pull-up handle, AM, bat **$30.00**

4-C-40, horizontal, plastic, step-down top with thumbwheel tuning knob, lower front on/off/volume knob, right and left circular grill areas with horizontal bars, twin speakers, swing handle, AM, bat **$75.00**

Fuji Denki

TRB-611, vertical, 1962, six transistors, upper right front thumb-wheel dial, lower perforated grill area with lower left logo, AM, bat **$40.00**

TRS-701, horizontal, 1962, upper front horizontal slide rule dial with thumbwheel tuning, large lower perforated grill area with MW/SW switch, telescoping antenna, AM, SW, bat **$30.00**

Futura

222, radio/phono, 1962, six transistors, lift top radio/phono with inner three-speed turntable and radio, handle, AM, bat **$35.00**

250, radio/phono, 1962, six transistors, lift top radio/phono with inner three-speed turntable and radio, handle, AM, bat **$35.00**

366, vertical, 1963, six transistors, upper right front thumbwheel dial, lower perforated grill area, AM, bat **$30.00**

Gala

TR-824, vertical, 1965, eight transistors, upper front window dial with thumbwheel tuning, large lower perforated grill area, AM, bat **$20.00**

Galaxie

FT-881, vertical, 4¼x2⅝x1¼", plastic, eight transistors, upper front window dial with right side thumbwheel tuning, left side thumbwheel on/off/volume knob, lower perforated grill area, made in Ryukyu, AM, bat $25.00

General Electric

7-2753D "76," vertical/Bicentennial radio, 4x2½x1¼", 1976, red, white, and blue plastic, upper right front window dial with right side thumbwheel tuning, left side thumbwheel on/off/volume knob, metal perforated grill painted with red, white, and blue "76," braided strap, made in Hong Kong, AM, bat **$25.00**

675, horizontal, 3⅛x5½x1½", 1956, black plastic, five transistors, GE's first commercially produced transistor radio, right front round dial with brass knob, top thumbwheel on/off/volume knob, left checkered grill area, made in USA, AM, bat **$100.00**

676, horizontal, 3⅛x5⅛x1½", 1956, ivory plastic, five transistors, GE's first commercially produced transistor radio, right front round dial with brass knob, top thumbwheel on/off/volume knob, left checkered grill area, made in USA, AM, bat **$100.00**

677, horizontal, 3⅛x5½x1½", 1956, red plastic, five transistors, GE's first commercially produced transistor radio, right front round dial with brass knob, top thumbwheel on/off/volume knob, left checkered grill area, made in USA, AM, bat $125.00

678, horizontal, 3⅛x5½x1½", 1956, aqua plastic, five transistors, GE's first commercially produced transistor radio, right front round dial with brass knob, top thumbwheel on/off/volume knob, left checkered grill area, made in USA, AM, bat .. $125.00

C2418A "Mickey Mouse," horizontal/clock radio, 6x10¾x5¼", white plastic, right front round dial with 3-D Mickey head pointer over horizontal bars, left alarm clock face with other Disney characters, AM, AC $50.00

C2419A "Mickey Mouse," horizontal/clock radio, 6x10¾x5¼", yellow plastic, right front round dial with 3-D Mickey head pointer over horizontal bars, left alarm clock face with other Disney characters, AM, AC **$50.00**

CT455A, horizontal/clock radio, 1960, six transistors, right front round dial knob over large perforated grill area, left alarm clock face, AM, bat **$20.00**

KT-1P-2751C, vertical, 4x2½x1¼", plastic, upper right front window dial with right side thumbwheel tuning, left side thumbwheel on/off/volume knob, front perforated grill area and right vertical panel with six stars, braided strap, made in Hong, Kong, AM, bat **$20.00**

P710A, horizontal, 1958, four transistors, right front round dial knob with "bubble" magnifier, center on/off/volume knob, left lattice grill area, AM, bat **$40.00**

P711A, horizontal, 3½x6⅜x1⅝", 1957, plastic, four transistors, right front round dial knob with "bubble" magnifier, center on/off/volume knob, left lattice grill area, AM, bat........ **$40.00**

P715-D, vertical, 6¾x3½x1¼", 1958, metal and leatherette, six transistors, upper front round dial, center on/off/volume knob, lower metal perforated grill area, pull-up handle, AM, bat $35.00

P725A, horizontal, 1958, tan plastic with brown handle, six transistors, right side dial knob, left side on/off/volume knob, large front perforated grill area, fold-down handle, AM, bat **$25.00**

P725B, horizontal, 1958, plastic, six transistors, right side dial knob, left side on/off/volume knob, large front perforated grill area, fold-down handle, AM, bat..................... **$25.00**

P-726A, horizontal, 6x9¼x2½", 1958, plastic with white handle, six transistors, right side dial knob, left side on/off/volume knob, large front metal perforated grill area, fold-down handle, made in USA, AM, bat $25.00

P-726B, horizontal, 6x9¼x2½", 1958, plastic with white handle, six transistors, right side dial knob, left side on/off/volume knob, large front metal perforated grill area, fold-down handle, made in USA, AM, bat **$25.00**

P740A, vertical, 1965, eight transistors, upper right front window dial with thumbwheel tuning, lower metal perforated grill area, AM, bat **$15.00**

P745B, horizontal, 3½x6⅝x1⅞", 1959, ebony plastic, five transistors, right front round dial knob, upper right thumbwheel on/off/volume knob, left grill area with vertical bars, made in USA, AM, bat **$25.00**

P746A, horizontal, 1958, plastic, five transistors, right front round dial knob, upper right thumbwheel on/off/volume knob, left grill area with vertical bars, AM, bat **$25.00**

P746B, horizontal, 3½x6⅝x1⅞", 1959, white and turquoise plastic, five transistors, right front round dial knob, upper right thumbwheel on/off/volume knob, left grill area with vertical bars, made in USA, AM, bat $25.00

P750A, horizontal, 1958, leather, six transistors, right side dial knob, left side on/off/volume knob, plastic front lattice grill area, leather handle, AM, bat $25.00

P755A, horizontal, 1959, plastic, five transistors, upper right front round dial knob, left thumbwheel on/off/volume knob, random pattern metal perforated grill area, pull-up handle, AM, bat **$20.00**

P760A, horizontal, 7¼x9⅜x2¾", 1958, beige and white plastic, five transistors, right side dial knob, left side on/off/volume knob, front lattice grill area, handle, AM, bat **$20.00**

P761A, horizontal, 7¼x9⅜x2¾", 1958, green and white plastic, five transistors, right side dial knob, left side on/off/volume knob, front lattice grill area, handle, AM, bat **$20.00**

P765A, vertical, 1957, metal and leatherette, six transistors, upper front round dial, center on/off/volume knob, lower metal perforated grill area, pull-up handle, rechargeable, AM, bat **$35.00**

P766A, vertical, 6¾x3½x1⅜", 1958, metal and leatherette, six transistors, upper front round dial, center on/off/volume knob, lower metal perforated grill area, pull-up handle, rechargeable, AM, bat $35.00

P-770, horizontal, 6½x8⅞x3⅛", 1959, plastic, seven transistors, upper right front round dial, upper left thumbwheel on/off/volume knob, large grill area, handle, AM, bat **$25.00**

P776A, horizontal, 1959, leather, seven transistors, upper right front round dial, upper left on/off/volume knob, lower grill area with horizontal bars, top right pushbutton, leather handle, AM, bat **$20.00**

P776B, horizontal, 6½x9½x3", 1959, leather, seven transistors, upper right front round dial, upper left on/off/volume knob, lower plastic grill area with horizontal bars, top right pushbutton, leather handle, AM, bat $20.00

P780A, horizontal, 1960, plastic and chrome, eight transistors, upper front horizontal slide rule dial, two knobs, large lower lattice grill area, leather handle, AM, bat **$20.00**

P780B, horizontal, 1960, plastic and chrome, eight transistors, upper front horizontal slide rule dial, two knobs, large lower lattice grill area, leather handle, AM, bat **$20.00**

P780H "Long Range," horizontal, 1965, plastic and chrome, nine transistors, upper front horizontal slide rule dial, two knobs, large lower lattice grill area, leather handle, AM, bat **$20.00**

P787A, horizontal, 1960, seven transistors, upper right front horizontal slide rule dial with right side thumbwheel tuning, lower right side on/off/volume knob, large grill area with circular cut-outs, AM, bat **$30.00**

P790A, horizontal, 1960, plastic, six transistors, right front round dial knob, lower right thumbwheel on/off/volume knob, left grill area with circular cut-outs, right side curves in, AM, bat **$35.00**

P790B, horizontal, 3¼x5⅝x1½", 1960, white and black plastic, six transistors, right front round dial knob, lower right thumbwheel on/off/volume knob, left grill area with circular cut-outs, right side curves in, made in USA, AM, bat $35.00

P791A, horizontal, 1960, plastic, six transistors, right front round dial knob, lower right thumbwheel on/off/volume knob, left grill area with circular cut-outs, right side curves in, AM, bat **$35.00**

P791B, horizontal, 3¼x5⅝x1½", 1960, turquoise and white plastic, six transistors, right front round dial knob, lower right thumbwheel on/off/volume knob, left grill area with circular cut-outs, right side curves in, made in USA, AM, bat **$35.00**

P-795A, horizontal, 1958, black leather, right side dial knob, left side on/off/volume knob, plastic front lattice grill area, leather handle, AM, bat **$25.00**

P-795B, horizontal, 1958, leather, right side dial knob, left side on/off/volume knob, plastic front lattice grill area, leather handle, AM, bat **$25.00**

P-796A, horizontal, 1958, blue leather, right side dial knob, left side on/off/volume knob, plastic front lattice grill area, leather handle, AM, bat................. **$25.00**

P-797A, horizontal, 1958, beige leather, right side dial knob, left side on/off/volume knob, plastic front lattice grill area, leather handle, AM, bat................. **$25.00**

P798C, horizontal, 1962, leather, right side dial knob, left side on/off/volume knob, plastic front lattice grill area, leather handle, AM, bat **$25.00**

P800A, horizontal, 3½x6¼x1¾", 1959, five transistors, right front round dial knob with "bubble" magnifier, center on/off/volume knob, left lattice grill area, made in USA, AM, bat **$30.00**

P805A, horizontal, 4¾x7x2¼", 1959, white plastic, five transistors, upper right front round dial knob, left thumbwheel on/off/volume knob, random pattern metal perforated grill area, pull-up handle, made in USA, AM, bat $20.00

P806A, horizontal, 4¾x7x2¼", 1959, blue plastic, five transistors, upper right front round dial knob, left thumbwheel on/off/volume knob, random pattern metal perforated grill area, pull-up handle, AM, bat **$20.00**

P807A, horizontal, 4¾x7x2¼", 1962, plastic, five transistors, upper right front round dial knob, left thumbwheel on/off/volume knob, woven grill area, pull-up handle, AM, bat................................. **$20.00**

P807B, horizontal, 4¾x7x2¼", 1962, plastic, five transistors, upper right front round dial knob, left thumbwheel on/off/volume knob, woven grill area, pull-up handle, AM, bat................................. **$20.00**

P807C, horizontal, 4¾x7x2¼", 1962, black plastic, five transistors, upper right front round dial knob, left thumbwheel on/off/volume knob, woven grill area, pull-up handle, AM, bat................................. **$20.00**

P807E, horizontal, 4¾x7x2¼", 1962, black plastic, five transistors, upper right front round dial knob, left thumbwheel on/off/volume knob, woven grill area, pull-up handle, AM, bat $20.00

P807H, horizontal, 4¾x7x2¼", 1962, black plastic, five transistors, upper right front round dial knob, left thumbwheel on/off/volume knob, woven grill area, pull-up handle, AM, bat **$20.00**

P-807J, horizontal, 4¾x7x2¼", 1962, black plastic, five transistors, upper right front round dial knob, left thumbwheel on/off/volume knob, woven grill area, pull-up handle, AM, bat **$20.00**

P-808A, horizontal, 4¾x7x2¼", 1959, plastic, upper right front round dial knob, left thumbwheel on/off/volume knob, woven grill area, pull-up handle, AM, bat **$20.00**

P-808C, horizontal, 4¾x7x2¼", 1962, white plastic, five transistors, upper right front round dial knob, left thumbwheel on/off/volume knob, woven grill area, pull-up handle, AM, bat **$20.00**

P808E, horizontal, 4¾x7x2¼", 1962, white plastic, five transistors, upper right front round dial knob, left thumbwheel on/off/volume knob, woven grill area, pull-up handle, AM, bat **$20.00**

P-808H, horizontal, 4¾x7x2¼", 1962, white plastic, five transistors, upper right front round dial knob, left thumbwheel on/off/volume knob, woven grill area, pull-up handle, AM, bat **$20.00**

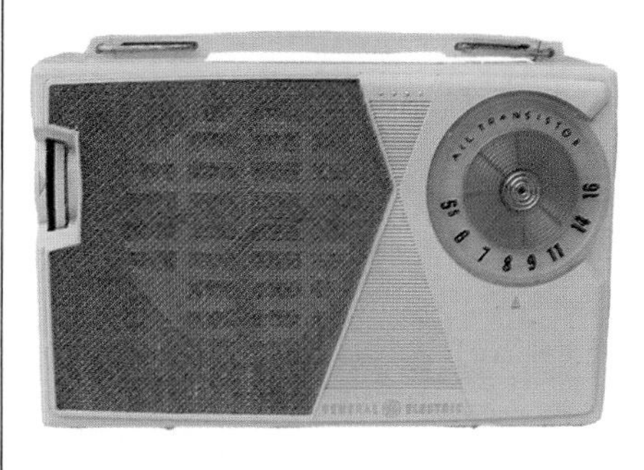

P-808J, horizontal, 4¾x7x2¼", 1962, white plastic, five transistors, upper right front round dial knob, left thumbwheel on/off/volume knob, woven grill area, pull-up handle, AM, bat $20.00

P-809C, horizontal, 4¾x7x2¼", 1962, green plastic, five transistors, upper right front round dial knob, left thumbwheel on/off/volume knob, woven grill area, pull-up handle, AM, bat **$20.00**

P-809E, horizontal, 4¾x7x2¼", 1962, green plastic, five transistors, upper right front round dial knob, left thumbwheel on/off/volume knob, woven grill area, pull-up handle, AM, bat **$20.00**

P-810A, horizontal, 6¾x8x2⅜", 1963, nutmeg leather and chrome, five transistors, right front dial, left thumbwheel on/off/volume knob, large metal perforated grill area, metal handle, made in USA, AM, bat $15.00

P-811A, horizontal, 6¾x8x2⅜", 1963, pearl white leather and chrome, five transistors, right front dial, left thumbwheel on/off/volume knob, large metal perforated grill area, metal handle, AM, bat $15.00

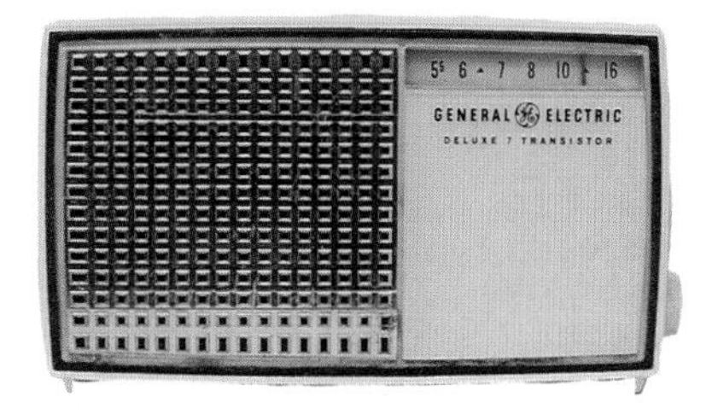

P815A, horizontal, 3⅜x6x1⅞", 1961, plastic, seven transistors, upper right front horizontal slide rule dial with right side thumbwheel tuning, lower right side on/off/volume knob, checkered grill area, AM, bat $25.00

P-820A, vertical, 4⅝x2⅞x1⅜", 1963, black and white plastic, six transistors, upper front horizontal slide rule dial with right side thumbwheel tuning, right side thumbwheel on/off/volume knob, lattice grill area, made in USA, AM, bat $20.00

P821A, vertical, 4⅝x2⅞x1⅜", 1963, blue and white plastic, six transistors, upper front horizontal slide rule dial with right side thumbwheel tuning, right side thumbwheel on/off/volume knob, lattice grill area, made in USA, AM, bat $20.00

P825A, vertical, 1961, upper right front round dial with right side thumbwheel tuning, upper left front thumbwheel on/off/volume knob, lower perforated grill area, handle, made in USA, AM, bat $30.00

P830A, vertical, 1960, charcoal, six transistors, upper front horizontal slide rule dial with right side thumbwheel tuning, lower perforated grill area, AM, bat $25.00

P830E, vertical, 1960, charcoal, six transistors, upper front horizontal slide rule dial with right side thumbwheel tuning, lower perforated grill area, AM, bat $25.00

P831A, vertical, 1960, blue, six transistors, upper front horizontal slide rule dial with right side thumbwheel tuning, lower perforated grill area, AM, bat $25.00

P840A, horizontal, 1963, leather, seven transistors, upper right front dial, lower on/off/volume knob, left grill area with oval cut-outs, leather handle, AM, bat..................... $25.00

P845A, horizontal, 1963, eight transistors, slanted top, right front vertical dial with thumbwheel tuning, thumbwheel on/off/volume knob, perforated grill area, handle, AM, bat $20.00

P850B, vertical, 3½x2⅜x1", 1962, plastic and metal, six transistors, upper right front round dial with right side thumbwheel tuning, left front thumbwheel on/off/volume knob, lower perforated grill area, top ring handle, AM, bat $30.00

P-850C, vertical, 3½x2⅜x1", 1962, black plastic and metal, six transistors, upper right front round dial with right side thumbwheel tuning, left front thumbwheel on/off/volume knob, lower perforated grill area, top ring handle, made in USA, AM, bat **$30.00**

P-850D, vertical, 3½x2⅜x1", 1964, black plastic and metal, six transistors, upper right front round dial with right side thumbwheel tuning, left front thumbwheel on/off/volume knob, lower perforated grill area, top ring handle, made in USA, AM, bat $30.00

P850E, vertical, 3½x2⅜x1", 1964, plastic and metal, six transistors, upper right front round dial with right side thumbwheel tuning, left front thumbwheel on/off/volume knob, lower perforated grill area, top ring handle, AM, bat **$30.00**

P-851C, vertical, 3½x2⅜x1", 1962, white plastic and metal, six transistors, upper right front round dial with right side thumbwheel tuning, left front thumbwheel on/off/volume knob, lower perforated grill area, top ring handle, made in USA, AM, bat **$30.00**

P-851D, vertical, 3½x2⅜x1", 1964, white plastic and metal, six transistors, upper right front round dial with right side thumbwheel tuning, left front thumbwheel on/off/volume knob, lower perforated grill area, top ring handle, AM, bat **$30.00**

P855A, horizontal, 1964, eight transistors, right front round dial, upper left thumbwheel on/off/volume knob, perforated grill area, fold-down handle, AM, bat **$15.00**

P856A, horizontal, 1964, eight transistors, right front round dial, upper left thumbwheel on/off/volume knob, perforated grill area, fold-down handle, made in USA, AM, bat **$15.00**

P-865A, horizontal, 1962, 11 transistors, upper front horizontal two-band slide rule dial, two knobs, lower lattice perforated grill area, telescoping antenna, AM, FM, bat **$25.00**

P870A, horizontal, 1962, eight transistors, comes with bottom clip-stand for use as a portable or car radio, front double-sided horizontal slide rule dial, two knobs, top perforated grill area, telescoping antenna, strap, AM, bat **$25.00**

P-875A, horizontal, 6¼x8½x3½", 1963, leather, eight transistors, upper right front dial, lower on/off/volume knob with bass/treble switch, left chrome grill area with oval cut-outs, pushbutton dial light, leather handle, made in USA, AM, bat $25.00

P-875D, horizontal, 6¼x8½x3½", 1963, leather, eight transistors, upper right front dial, lower on/off/volume knob with bass/treble switch, left chrome grill area with oval cut-outs, pushbutton dial light, leather handle, made in USA, AM, bat **$25.00**

P885B, vertical, 1963, six transistors, upper right front round dial knob, upper left front thumbwheel on/off/volume knob, lower grill area with horizontal bars, AM, bat **$15.00**

P891A, horizontal, 1965, eight transistors, upper right front dial, two upper left front knobs, large grill area, handle, AM, bat **$10.00**

P-916E, vertical, 4¾x3⅛x1⅜", plastic, eight transistors, upper right front round dial knob, left side thumbwheel on/off/volume knob, lower metal perforated grill area, AM, bat $15.00

P920A "Long Range," horizontal, 4⅞x8x2¾", 1965, leather, 10 transistors, center front vertical slide rule dial, two right thumbwheel knobs and one tone switch, left metal perforated grill area, leather handle, AM, bat **$15.00**

P-920B "Long Range," horizontal, 4⅞x8x2¾", 1965, leather, 10 transistors, center front vertical slide rule dial, two right thumbwheel knobs and one tone switch, left metal perforated grill area, leather handle, AM, bat **$15.00**

P930A, horizontal, 1963, eight transistors, upper left front horizontal three-band slide rule dial, large lower perforated grill area, telescoping antenna, handle, AM, 2SW, bat $25.00

P940C, horizontal, 5¼x8½x2½", 1965, 13 transistors, right front two-band dial, two right side knobs, large front perforated grill area with AM/FM and tone switches, telescoping antenna, handle, AM, FM, bat $15.00

P943C, horizontal, 1965, 14 transistors, right front two-band dial, two right side knobs, large front perforated grill area with AM/FM and AFC switches, telescoping antenna, handle, AM, FM, bat **$15.00**

P945B, (bottom left) vertical, 4½x3x 1¼", 1965, plastic, upper right front oval window dial with right side thumbwheel tuning, left side thumbwheel on/off/volume knob, lower horizontal grill bars, AM, bat .. $10.00

P955A, horizontal, 1963, seven transistors, upper front horizontal slide rule dial, two knobs, hi/lo switch, large lower perforated grill area, handle, AM, bat **$15.00**

P955B, vertical, 1965, seven transistors, upper right front round dial knob, lower perforated grill area, AM, bat **$15.00**

P965A, vertical, 1965, leather, 10 transistors, upper front horizontal two-band slide rule dial, lower perforated grill area, telescoping antenna, leather handle, AM, SW, bat .. **$15.00**

P968A, vertical, 1965, leather, 10 transistors, upper front horizontal two-band slide rule dial, lower perforated grill area, telescoping antenna, leather handle, AM, SW, bat .. **$15.00**

P970A, horizontal, 1964, 14 transistors, upper left front round two-band dial, three knobs, two switches, large lower perforated grill area, telescoping antenna, AM, FM, bat **$15.00**

P990A, horizontal, 1965, leather, 17 transistors, upper front horizontal five-band slide rule dial, large lower perforated grill area, telescoping antenna, leather handle, AM, FM, 2SW, LW, bat **$25.00**

P990C "World Monitor," horizontal, 8½x12x4¾", 1965, leather, 17 transistors, upper front horizontal five-band slide rule dial, large lower metal perforated grill area, two telescoping antennas, rear log pouch, leather handle, AM, FM, 2SW, LW, bat $25.00

P1700A, vertical, 1965, 10 transistors, upper right front window dial with thumbwheel tuning, lower metal perforated grill area, AM, bat **$15.00**

P1701A, (bottom left) vertical, 1965, plastic, 10 transistors, upper right front window dial with thumbwheel tuning, lower metal perforated grill area, AM, bat $15.00

P1710C, vertical, 4¼x2¾x1¼", 1965, plastic, upper front large round dial knob, right side thumbwheel on/off/volume knob, lattice grill area with lower left logo, AM, bat $10.00

P1730B, vertical, plastic, solid state, upper front round dial, lower metal perforated grill area, AM, bat **$15.00**

P-1757, vertical, 4x2½x1½", plastic, upper right front window dial with right side thumbwheel tuning, left side thumbwheel on/off/volume knob, lower horizontal grill bars, braided strap, made in Hong Kong, AM, bat **$10.00**

P-1758, vertical, 4x2½x1½", plastic, upper right front window dial with right side thumbwheel tuning, left side thumbwheel on/off/volume knob, lower horizontal grill bars, braided strap, made in Hong Kong, AM, bat $10.00

P1805A, horizontal, 4½x6¾x2⅛", plastic, nine transistors, upper front horizontal slide rule dial, lower metal perforated grill area, left side thumbwheel knob, handle, AM, bat $15.00

P1818B, horizontal, 5x7x2½", 1965, plastic, 10 transistors, upper right front round two-band dial, lower AM/FM switch, horizontal grill bars, fold-down handle, AM, FM, bat $10.00

P1821L, horizontal, 4½x6¾x2⅛", plastic, 11 transistors, upper front horizontal two-band slide rule dial, lower metal perforated grill area with right FM/AM switch and left logo, left side thumbwheel knob, telescoping antenna, handle, AM/FM, bat $15.00

P2790F, vertical, 4x2½x1⅜", plastic, solid state, upper right front window dial with right side thumbwheel tuning, left side thumbwheel on/off/volume knob, large grill area with horizontal bars, left side braided strap, made in Hong Kong, AM, bat **$10.00**

P-9001A, horizontal, 1962, seven transistors, upper right front horizontal slide rule dial with right side thumbwheel tuning, lower right side on/off/volume knob, left checkered grill area, AM, bat **$25.00**

P9011, horizontal, 1962, plastic, seven transistors, upper right front horizontal slide rule dial with right side thumbwheel tuning, lower right side on/off/volume knob, left checkered grill area, AM, bat $25.00

S-15, horizontal/kit radio, 4½x7½x 2⅜", 1963, sold as a kit, leather, upper right front round dial, left front thumbwheel on/off/volume knob, center grill area with circular cut-outs, handle, AM, bat...... $15.00

Global

GFM-931, horizontal, 4x6¾x1½", plastic, 10 transistors, finished on both sides, top wrap-over see-through two-band dial, lower metal perforated grill area with logo, top left telescoping antenna, AM, FM, bat $40.00

GR-201 "Boy's Radio," horizontal, 2½x4x1¼", plastic, right front window dial with right side thumbwheel tuning, upper right front thumbwheel on/off/volume knob, left grill area with horizontal bars, AM, bat $35.00

GR-711, vertical, 3¾x2¾x1", plastic, six transistors, upper front V-shaped window dial with right side thumbwheel tuning, left side thumbwheel on/off/volume knob, metal perforated grill area, swing handle, AM, bat $35.00

GR-823, horizontal, 2⅞x7⅞x1¾", plastic, eight transistors, two upper front horizontal slide rule dials – one AM, one SW – lower metal perforated grill area with logo, BC/SW switch, made in Japan, AM, SW, bat............. $25.00

GR-900, vertical, 4½x3x1¼", 1963, plastic, nine transistors, upper front half-round dial with right side thumbwheel tuning, right side thumbwheel on/off/volume knob, metal perforated grill area with logo, AM, bat $45.00

GR-920, horizontal, 1965, 10 transistors, upper front horizontal two-band slide rule dial, right side thumbwheel knobs, horizontal grill bars, telescoping antenna, AM, FM, bat **$15.00**

Global Imperial

HT-8054, vertical, 4¼x2½x1¼", plastic, eight transistors, metal front, upper right window dial with right side thumbwheel tuning, top left thumbwheel on/off/volume knob, perforated grill area, AM, bat $25.00

Gloria

"Boy's Radio," vertical, 3¾x2½x 1¼", plastic, two transistors, upper right front window dial with right side thumbwheel tuning, left side thumbwheel on/off/volume knob, round grill area with segmented horizontal bars, AM, bat....... **$30.00**

Golden Shield

7040, horizontal, 3½x6x1½", plastic, upper front horizontal two-band slide rule dial, upper right front thumbwheel dial, upper left front thumbwheel on/off/volume knob, large perforated grill area with gold logo, telescoping antenna, AM, Marine, bat $35.00

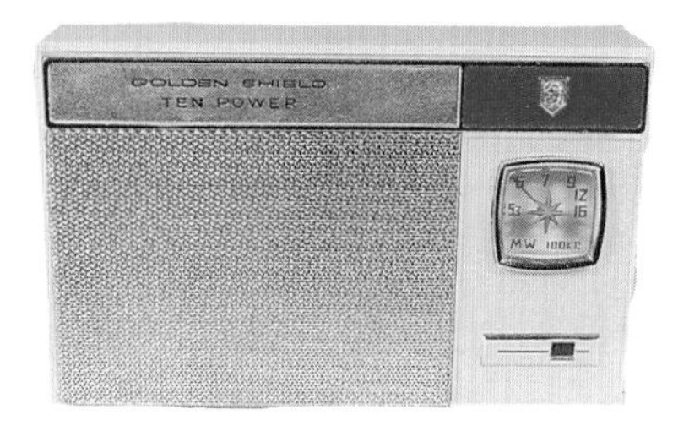

7109 "Ten Power," horizontal, 3¾x 6⅛x1¾", plastic, eight transistors, right front dial with right side thumbwheel tuning, lower right front on/off/volume window with right side thumbwheel knob, large left metal perforated grill area, shield logo, made in Japan, AM, bat $35.00

7186, vertical, plastic, six transistors, step-back top, upper left front window dial over large grill area with horizontal bars, upper left thumbwheel tuning, right side thumbwheel on/off/volume knob, AM, bat $15.00

7309 "Personal," vertical, plastic, six transistors, upper right front window dial with right side thumbwheel tuning, upper left front golden shield logo, lower metal perforated grill area, AM, bat **$25.00**

Grundig

Mini-Boy Transistor 200, horizontal, 2½x4⅛x1", 1964, plastic, six transistors, front diagonally divided with brass trim, right side see-through window dial with thumbwheel tuning, lower right front thumbwheel on/off/volume knob, left painted perforated grill area, rear fold-out stand, sold with matching speaker

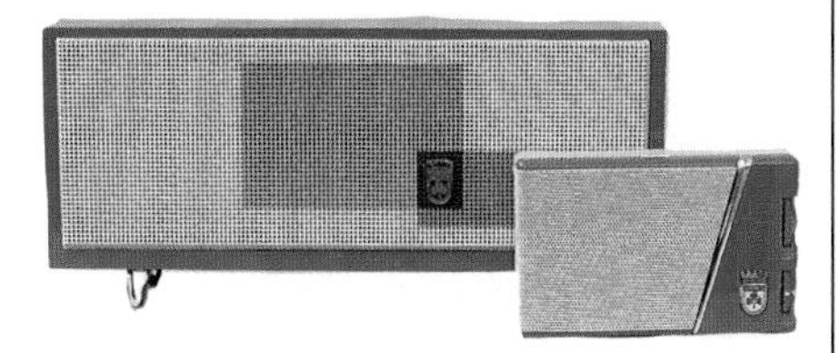

box, made in West Germany, AM, bat.
radio without speaker box $35.00
radio with speaker box $65.00

Ocean-Boy 204, horizontal, 1965, 17 transistors, right front vertical six-band slide rule dial and nine pushbuttons, left grill area with horizontal bars, telescoping antenna, handle, LW, MW, 3SW, FM, bat **$35.00**

Prima-Boy 201E, horizontal, 1963, nine transistors, top wrap-over three-band dial, large lower grill area with horizontal bars, thumbwheel knobs, telescoping antenna, handle, AM, FM, SW, bat **$25.00**

Teddy-Boy, horizontal, 1961, eight transistors, rounded case with top three band slide rule dial, pushbuttons, and two knobs, large front lattice grill area, two telescoping antennas, AM, FM, SW, bat **$30.00**

Transonette 89 "Black Magic Standard," horizontal/table, 1963, nine transistors, lower front horizontal two-band slide rule dial, large upper lattice grill area with three right pushbuttons, two telescoping antennas, AM, FM, bat $20.00

Transonette 99U, horizontal/table, 1964, nine transistors, lower front horizontal four-band slide rule dial, large upper lattice grill area with seven right pushbuttons, telescoping antenna, AM, FM, LW, SW, bat **$20.00**

Transworld Junior, horizontal, 1961, eight transistors, upper right front round three-band dial overlaps horizontal grill bars, thumbwheel knobs, three pushbuttons, telescoping antenna, handle, AM, FM, SW, bat **$20.00**

Transworld TR16, horizontal, 1964, 16 transistors, right front vertical six-band slide rule dial and nine pushbuttons, left grill area with horizontal bars, telescoping antenna, handle, LW, MW, 3SW, FM, bat **$30.00**

Hallicrafters

TR-88, horizontal, 1958, leather, six transistors, upper left front dial knob, upper right front on/off/volume knob, lower grill area with cut-outs, leather handle, AM, bat **$30.00**

Hamilton

YT-161, vertical, 4x2½x1¼", plastic, six transistors, upper right thumbwheel dial knob, left on/off/volume window with left side thumbwheel knob, lower metal perforated grill area, AM, bat **$25.00**

Harpers

2TP-110, vertical, 3¼x2¼x1¼", plastic, large front round dial, top on/off/switch, no speaker, earphone is wired in, no volume control, AM, bat $75.00

547F, horizontal, plastic, diagonally divided front with right round dial and left horizontal grill bars, lower perforated horizontal panel, AM, bat **$35.00**

GK-200, horizontal, 1962, eight transistors, upper front horizontal two-band slide rule dial with thumbwheel tuning, left on/off/volume window, lower perforated grill area, top pushbutton, telescoping antenna, AM, SW, bat **$30.00**

GK-631, vertical, 1962, six transistors, upper front V-shaped window dial with right side thumbwheel tuning, left side thumbwheel on/off/volume knob, lower grill area with horizontal slots, AM, bat **$35.00**

Heathkit

GR-151A, horizontal, 6½x8⅛x3¼", leather, six transistors, lower right front dial knob, lower left front on/off/volume knob, large grill area with circular cut-outs, leather handle, AM, bat $10.00

XR-2L, horizontal, 1960, leather and plastic, six transistors, kit radio, wedge-shaped case with side knobs, large front grill area with lower right logo, handle, AM, bat **$10.00**

XR-2P, horizontal, 1960, plastic, six transistors, kit radio, wedge-shaped case with side knobs, large front grill area with lower right logo, handle, AM, bat **$10.00**

Hemisphere

AM-6T4, vertical, 1964, six transistors, upper right front window dial with right side thumbwheel tuning, left side thumbwheel on/off/volume knob, perforated grill area, AM, bat **$15.00**

AM-8T4, vertical, 1964, eight transistors, upper right front window dial with right side thumbwheel tuning, upper left front on/off/volume window with left side thumbwheel knob, perforated grill area, AM, bat **$15.00**

Hi-Delity

7TA-1X, horizontal, 1964, eight transistors, right front two-band window dial with right side thumbwheel tuning, lower right thumbwheel on/off/volume knob, left grill area with horizontal bars, telescoping antenna, AM, SW, bat **$45.00**

7TA-1Y, horizontal, 1964, eight transistors, right front two-band window dial with right side thumbwheel tuning, lower right thumbwheel on/off/volume knob, left grill area with horizontal bars, telescoping antenna, AM, SW, bat **$45.00**

CFM-3-1200S, horizontal, 1964, 10 transistors, upper front horizontal three-band slide rule dial, large lower grill area with right and left switches, telescoping antenna, AM, FM, SW, bat **$25.00**

CFM-1200S, horizontal, 1964, 10 transistors, upper front horizontal two-band slide rule dial, large lower grill area with right and left switches, telescoping antenna, AM, FM, bat **$25.00**

N-601, horizontal, 1963, six transistors, upper right front window dial with right side thumbwheel tuning, lower right side thumbwheel on/off/volume knob, large grill area with vertical slots, AM, bat **$20.00**

N-801, horizontal, 1964, eight transistors, upper right front window dial with right side thumbwheel tuning, lower right side thumbwheel on/off/volume knob, large grill area with vertical slots, AM, bat **$20.00**

SR-H600 "HiFidelity," horizontal, 1964, eight transistors, upper front horizontal four-band slide rule dial, lower perforated grill area, telescoping antenna, handle, AM, 3SW, bat **$25.00**

STH-601, horizontal, 1963, six transistors, diagonally divided front with right globe dial and left grill area, two knobs, base, handle, AM, bat **$25.00**

Hi-Lite

STW-6, horizontal/clock radio, 1964, six transistors, lower front horizontal slide rule dial under center clock face, right and left grill areas, telescoping antenna, AM, bat **$15.00**

Hilton

TR8A7 "High Fidelity," vertical, 4⅜x2¾x1⅜", plastic, eight transistors, upper left front round dial knob over large metal perforated grill area, right side thumbwheel on/off/volume knob, made in Japan, AM, bat **$25.00**

TR108, vertical, 4⅜x2¾x1⅜", plastic, 10 transistors, upper left front round dial knob over large metal perforated grill area with lower right logo, right side thumbwheel on/off/volume knob, made in Japan, AM, bat **$25.00**

Hitachi

KH-903, horizontal, 1964, nine transistors, upper front horizontal two-band slide rule dial, two knobs, large lower grill area with horizontal bars, telescoping antenna, handle, AM, FM, bat........ **$20.00**

KH-915, horizontal, 1963, nine transistors, right front FM window dial with right side thumbwheel tuning, lower right side thumbwheel on/off/volume knob, left circular checkered grill area, telescoping antenna, FM, bat.................. **$25.00**

KH-960H "Hiphonic," horizontal, 1965, available in black or beige, nine transistors, top horizontal two-band slide rule dial, large front perforated grill area with lower left logo, telescoping antenna, AM, FM, bat.......... **$20.00**

KH-1000H "Hi-Phonic," horizontal, 1965, 10 transistors, top horizontal two-band slide rule dial, front grill area with horizontal slots, telescoping antenna, AM, FM, bat...... **$20.00**

KH-1005, horizontal, 1963, 10 transistors, upper front horizontal two-band slide rule dial, lower grill area with horizontal bars and lower right logo, four top pushbuttons, telescoping antenna, handle, AM, FM, bat....................................... **$25.00**

T-728, horizontal/clock radio, 1963, seven transistors, right front dial, left alarm clock face, center perforated grill area with logo, AM, bat .. **$10.00**

TH-600, vertical, 1964, plastic, six transistors, upper left front window dial, right side thumbwheel on/off/volume knob, large round metal perforated grill area with movable outer ring used for tuning, top vinyl strap, AM, bat............. $35.00

TH-621, horizontal, 1959, six transistors, right front round dial knob, lower right front thumbwheel on/off/volume knob, left lattice grill area, right side strap, AM, bat **$30.00**

TH-627R, vertical, 1961, six transistors, upper front window dial with right side thumbwheel tuning, large lower perforated grill area with center logo, AM, bat.................... **$20.00**

TH-640, vertical, 1964, six transistors, upper front window dial with right side thumbwheel tuning, large lower perforated grill area with lower right logo, AM, bat **$15.00**

TH-650, vertical, 1963, six transistors, upper front window dial with right side thumbwheel tuning, large lower perforated grill area, AM, bat **$15.00**

TH-660, vertical, 1962, six transistors, window dial with right side thumbwheel tuning, right side thumbwheel on/off/volume knob, large front perforated grill area with center logo, AM, bat **$30.00**

TH-666, vertical, 4x2¼x1¼", 1959, available in red/gray, gold/black, or pearl/white plastic, six transistors, upper right front dial with right side thumbwheel tuning, upper left front volume window with left side thumbwheel on/off/volume knob, metal perforated grill area, could be used with optional speaker box model #ES-90H, AM, bat.
radio without speaker box $75.00
radio with speaker box $115.00

TH-667, horizontal, 4¼x6¾x1¾", 1960, available in pale pink, light blue, or mint green plastic, six transistors, recessed right front with thumbwheel tuning and on/off/volume knobs, large checkered front grill area with upper left logo, made in Japan, AM, bat $40.00

TH-680 "Hi-Phonic," vertical, 1965, six transistors, upper front round dial, left side thumbwheel on/off/volume knob, large perforated grill area, AM, bat **$20.00**

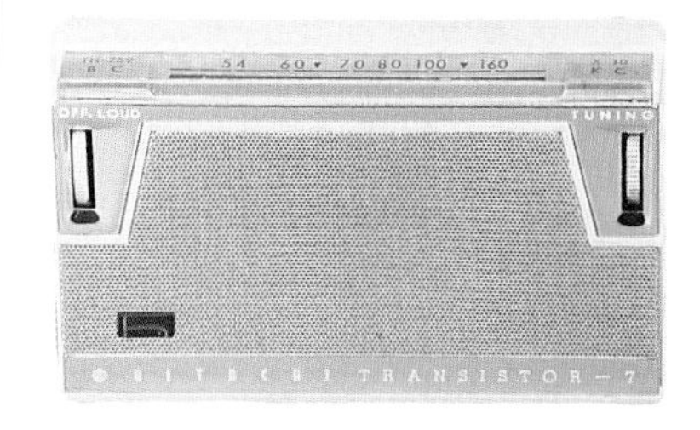

TH-759, horizontal, 1962, seven transistors, top horizontal slide rule dial with right front thumbwheel tuning, left front thumbwheel on/off/volume knob, large metal perforated grill area with lower left logo, AM, bat $40.00

TH-812, horizontal, 1964, leather, eight transistors, upper front horizontal slide rule dial, large lower grill area with right tuning and volume knobs, leather handle, AM, bat **$15.00**

TH-812R, horizontal, 6½x10x4½", 1964, leather, eight transistors, upper front horizontal slide rule dial, large lower grill area with right tuning and volume knobs, leather handle, AM, bat **$15.00**

TH-831, vertical, plastic, solid state, upper front horizontal slide rule dial, right and left side knobs, lower grill area with horizontal bars, top vinyl strap, AM, bat $20.00

TH-841, horizontal, 3¾x6¼x1½", 1964, plastic, eight transistors, top horizontal slide rule dial with right side thumbwheel tuning, lower right side thumbwheel on/off/volume knob, large front metal perforated grill area with lower right logo, AM, bat **$30.00**

TH-848, horizontal, 1964, available in black or beige, eight transistors, upper right front window dial with right side thumbwheel tuning, lower right front volume window with right side thumbwheel knob, large front grill area with lower left tone switch, AM, bat **$20.00**

TH-853, horizontal, 4x6⅞x1¾", available in ivory, red, green, or yellow plastic, solid state, top slanted slide rule dial with right side tuning knob and left side on/off/volume knob, front horizontal grill bars, right side AC plug-in, braided strap, made in Taiwan, AM, AC/bat $20.00

TH-862, horizontal, 2¾x4½x1¼", 1960, available in coral, black, or gray plastic, eight transistors, right front window dial with right side thumbwheel tuning, lower right front on/off/volume window with right side thumbwheel knob, large metal perforated grill area with upper right logo, AM, bat **$25.00**

TH-862R "Marie," horizontal, 2¾x4½x1¼", 1960, plastic, eight transistors, right front window dial with right side thumbwheel tuning, lower right front on/off/volume window with right side thumbwheel knob, large metal perfo-

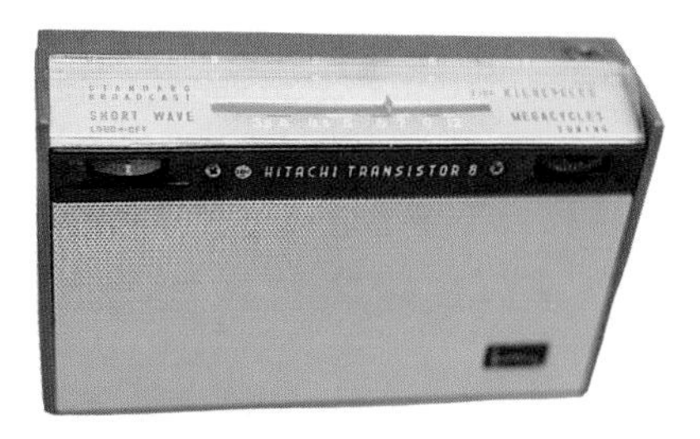

rated grill area with upper right logo, AM, bat $25.00

TH-890 "Hi-Phonic," horizontal, 1965, eight transistors, upper right front window dial with right side thumbwheel tuning, lower right front on/off/volume window with right side thumbwheel knob, large front perforated grill area, AM, bat **$15.00**

WH-761M, vertical, 1961, seven transistors, two upper front window dials - one Marine, one broadcast - with right side thumbwheel tuning, perforated grill area, AM, SW, bat **$25.00**

WH-761SB, vertical, 1962, seven transistors, two upper front window dials - one shortwave, one broadcast - with right side thumbwheel tuning, perforated grill area, AM, SW, bat **$25.00**

WH-817, horizontal, 1962, eight transistors, upper front horizontal three-band slide rule dial with thumbwheel tuning, lower perforated grill area with left "dial lamp" button, right band switch, telescoping antenna, AM, 2SW, bat **$30.00**

WH-822 "Kelly," horizontal, 3¾x6x 1¾", 1960, plastic, eight transistors, top horizontal two-band dial with upper right front thumbwheel tuning, upper left front thumbwheel on/off/volume knob, large metal perforated grill area with lower right logo, AM, SW, bat $30.00

WH-822H, horizontal, 3¾x6x1¾", 1964, plastic, eight transistors, top horizontal two-band dial with upper right front thumbwheel tuning, upper left front thumbwheel on/off/volume knob, large metal perforated grill area with lower right logo, AM, SW, bat **$30.00**

WH-822M, horizontal, 3¾x6x1¾", 1960, plastic, eight transistors, top horizontal two-band dial with upper right front thumbwheel tuning, upper left front thumbwheel on/off/volume knob, large metal perforated grill area with lower right logo, AM, Marine, bat **$30.00**

WH-822MB, horizontal, 3¾x6x1¾", 1960, two-tone gray plastic, eight transistors, top horizontal two-band dial with upper right front thumbwheel tuning, upper left front thumbwheel on/off/volume knob, large metal perforated grill area with lower right logo, AM, Marine, bat **$30.00**

WH-822SW, horizontal, 3¾x6x1¾", 1960, black plastic, eight transistors, top horizontal two-band dial with upper right front thumbwheel tuning, upper left front thumbwheel on/off/ volume knob, large metal perforated grill area with lower right logo, AM, SW, bat **$30.00**

WH-829, horizontal, 1962, eight transistors, top horizontal two-band slide rule dial with upper right front thumbwheel tuning, lower right front thumbwheel fine tuning knob, upper left front thumbwheel on/off/volume knob, large perforated grill area, telescoping antenna, AM, SW, bat **$30.00**

WH-829M, horizontal, 1963, eight transistors, top horizontal two-band slide rule dial with upper right front thumbwheel tuning, lower right front thumbwheel fine tuning knob, upper left front thumbwheel on/off/volume knob, large perforated grill area, telescoping antenna, AM, Marine, bat **$30.00**

WH-999 "Hiphonic," horizontal, 1965, nine transistors, upper front horizontal three-band slide rule dial, lower left lattice grill area and right band switch, light button, telescoping antenna, AM, 2SW, bat **$30.00**

XH-1500, horizontal, 6¼x9½x3", 1961, plastic, 15 transistors, upper front horizontal two-band illuminated slide rule dial, large lower perforated grill area, two top knobs and four pushbuttons, two telescoping antennas, handle, AM, FM, bat **$25.00**

Hoffman

709X "Solar," horizontal, 3x5x1¾", 1963, available in black, beige, or ivory plastic, nine transistors, upper front horizontal slide rule dial with right side thumbwheel tuning, lower right side thumbwheel on/off/volume knob, perforated grill area with lower left logo, top solar panel, AM, bat **$175.00**

719 "Solar," horizontal, 1963, seven transistors, right front round dial, large perforated grill area, top solar panel, AM, bat **$150.00**

727X, vertical, 1963, six transistors, upper front window dial with thumbwheel tuning, lower perforated grill area, AM, bat **$15.00**

729, horizontal, 1964, 12 transistors, upper front horizontal two-band slide rule dial, large lower perforated grill area, two telescoping antennas, handle, AM, FM, bat **$20.00**

759, horizontal, 1964, 14 transistors, upper front horizontal two-band slide rule dial, large lower perforated grill area, two telescoping antennas, handle, AM, FM, bat **$20.00**

BP706 "Trans Solar," horizontal, 3⅞x6¼x2", 1959, mocha plastic, six transistors, right front dial knob, left perforated grill area, top solar panel, swing handle, AM, bat **$175.00**

BP-709XS "Solar," horizontal, 3x5x1¾", plastic, nine transistors, upper front horizontal slide rule dial with

right side thumbwheel tuning, lower right side thumbwheel on/off/volume knob, metal perforated grill area with lower left logo, top solar panel, AM, bat $150.00

CP706 "Trans Solar," horizontal, 3⅞x 6¼x2", 1959, espresso plastic, six transistors, right front dial knob, left perforated grill area, top solar panel, swing handle, AM, bat **$175.00**

EP706 "Trans Solar," horizontal, 3⅞x 6¼x2", 1959, plastic, six transistors, right front dial knob, left perforated grill area, top solar panel, swing handle, AM, bat.................. **$175.00**

KP706 "Trans Solar," horizontal, 3⅞ x6¼x2", 1959, ebony plastic, six transistors, right front dial knob, left perforated grill area, top solar panel, swing handle, AM, bat....... $175.00

KP707, vertical, 1962, seven transistors, upper right front round dial knob with wedge-shaped opening, left side thumbwheel on/off/volume knob, lower perforated grill area with lower left logo, swing handle, AM, bat **$25.00**

KP-709X-CD "Solar," horizontal, 1964, plastic, nine transistors, upper front horizontal slide rule dial with right side thumbwheel tuning, lower right side thumbwheel on/off/volume knob, perforated grill area with lower left logo, top solar panel, AM, bat.................... **$150.00**

OP706 "Trans Solar," horizontal, 3⅞x6¼x2", 1959, oyster white plastic, six transistors, right front dial knob, left perforated grill area, top solar panel, swing handle, AM, bat............. **$175.00**

OP708, horizontal, 1962, eight transistors, upper front horizontal slide rule dial with right side thumbwheel tuning, lower right side thumbwheel on/off/volume knob, perforated grill area with lower left logo, AM, bat....................... **$30.00**

OP-709XS "Solar," horizontal, 1963, nine transistors, upper front horizontal slide rule dial with right side thumbwheel tuning, lower right side thumbwheel on/off/volume knob, perforated grill area with lower left logo, top solar panel, AM, bat.......... **$150.00**

PP706 "Trans Solar," horizontal, 3⅞x 6¼x2", 1959, pink plastic, six transistors, right front dial knob, left perforated grill area, top solar panel, swing handle, AM, bat **$175.00**

RP706 "Trans Solar," horizontal, 3⅞x 6¼x2", 1959, red plastic, six transistors, right front dial knob, left perforated grill area, top solar panel, swing handle, AM, bat **$175.00**

TP706 "Trans Solar," horizontal, 3⅞x 6¼x2", 1959, turquoise plastic, six transistors, right front dial knob, left perforated grill area, top solar panel, swing handle, AM, bat **$175.00**

Holiday

HS921 "Super DX," vertical, 4¼x2½x 1¼", plastic, nine transistors, upper right front window dial with right side thumbwheel tuning, left side thumbwheel on/off/volume knob, metal front panel with vertical grill slots, AM, bat $20.00

TS-190, horizontal, plastic, upper front horizontal multi-band slide rule dial, large lower metal perforated grill area, handle, bat **$15.00**

Honey Tone

8TP-412, vertical, 1963, eight transistors, upper front window dial with right side thumbwheel tuning, large lower perforated grill area with center logo, swing handle, AM, bat **$30.00**

604, vertical, 1963, six transistors, upper front horizontal slide rule dial with right side thumbwheel tuning, left side thumbwheel on/off/volume knob, large perforated grill area, AM, bat **$15.00**

FR-601, vertical, 1962, six transistors, upper front window dial with right side thumbwheel tuning, left side thumbwheel on/off/volume knob, round perforated grill area, swing handle, AM, bat **$30.00**

KTF-102G, horizontal, 1963, 11 transistors, upper front horizontal double slide rule dial, large lower perforated grill area with three knobs, telescoping antenna, handle, AM, FM, bat.......... **$15.00**

TR-801 "All Wave Super," horizontal, 1963, eight transistors, upper front horizontal two-band slide rule dial, lower right thumbwheel knobs, large perforated grill area, AM, SW, bat......................... **$25.00**

Hy-Lite

E164, vertical, 4⅝x2¾x1¼", plastic, solid state, upper right front round window dial with right side thumbwheel tuning, upper left front round on/off/volume window with left side thumbwheel knob, lower checkered grill area, made in Hong Kong, AM, bat $10.00

Imperial

GK-600 "Hi-Fi," vertical, 1963, six transistors, upper front window dial with side thumbwheel tuning, large lower grill area with horizontal slots, AM, bat **$20.00**

Invicta

8PK1, vertical, 1965, eight transistors, upper front window dial with side thumbwheel tuning, large lower grill area, AM, bat **$10.00**

10PK1, vertical, 1965, 10 transistors, upper front window dial with side thumbwheel tuning, large lower perforated grill area, AM, bat **$10.00**

ITT

600, vertical, 1963, six transistors, upper front horizontal slide rule dial with thumbwheel tuning, large lower perforated grill area with center "6" logo, AM, bat **$30.00**

615, vertical, 4x2⅝x¾", 1963, plastic, six transistors, upper left front window dial with top left thumbwheel tuning, top right thumbwheel on/off/volume knob, large metal perforated grill area with lower right logo, metal back, AM, bat $20.00

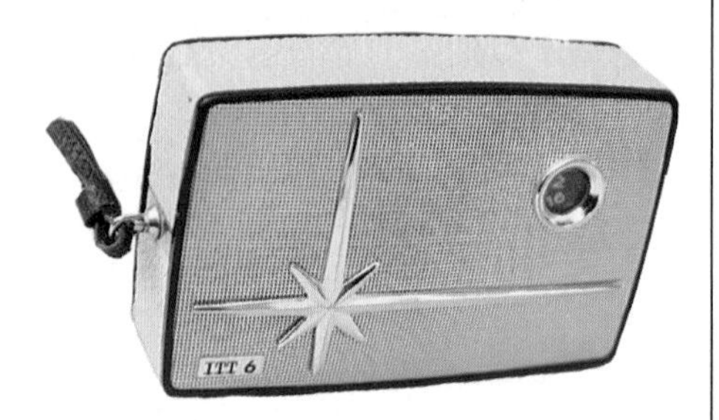

628, horizontal, 2¾x4¼x1", 1963, leatherette and metal, six transistors, right front round window dial with right side thumbwheel tuning, lower right side thumbwheel on/off/volume knob, large front metal perforated grill area with "star" emblem, left side strap, AM, bat $35.00

631, vertical, 1963, six transistors, upper front window dial with right side thumbwheel tuning, right side thumbwheel on/off/volume knob, perforated grill area, AM, bat **$20.00**

731, horizontal, 1963, seven transistors, upper right front horizontal dial with right side thumbwheel tuning, large perforated grill area with lower right logo, AM, bat **$30.00**

871, vertical, 1963, eight transistors, upper front horizontal slanted two-band slide rule dial with right side thumbwheel tuning, large lower perforated grill area, telescoping antenna, AM, SW, bat **$25.00**

881, horizontal, 3½x6¼x1⅜", 1963, eight transistors, upper front horizontal two-band slide rule dial with right side thumbwheel tuning, left side thumbwheel on/off/volume

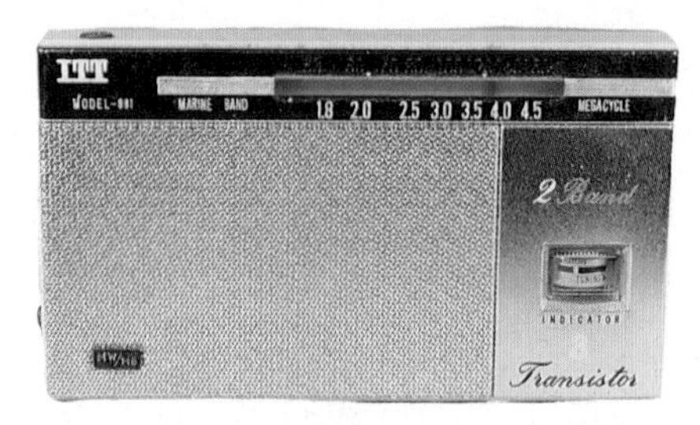

knob, large metal perforated grill, right battery/tuning indicator window, rear band switch, made in Japan, AM, Marine, bat $30.00

881-S, horizontal, 1963, eight transistors, upper front horizontal two-band slide rule dial with thumbwheel tuning, large perforated grill area, right battery/tuning indicator window, AM, SW, bat **$30.00**

1005, horizontal, 1963, nine transistors, upper front horizontal three-band slide rule dial, large lower perforated grill area, telescoping antenna, handle, AM, Marine, SW, bat **$20.00**

1011, horizontal, 1963, 10 transistors, upper front horizontal two-band slide rule dial, large lower perforated grill area with lower left logo, right and left side knobs, top pushbuttons, telescoping antenna, handle, AM, FM, bat **$25.00**

6409-A "Super-Sensitivity," vertical, 9x4½x1⅞", plastic, nine transistors, left front vertical slide rule dial, three front thumbwheel knobs, metal perforated grill area, handle, AM, bat **$15.00**

6409-F, horizontal, 1964, nine transistors, off-center vertical two-band dial with thumbwheel tuning, large front perforated grill area with lower left logo, telescoping antenna, handle, AM, FM, bat **$15.00**

6509, vertical, 1964, nine transistors, two upper front dials, one AM, one FM, lower perforated grill area with lower right logo, telescoping antenna, AM, FM, bat **$15.00**

J.C. Penney

620, horizontal, 1963, leather, 12 transistors, two upper front round dials - one AM, one FM, lower perforated grill area with left logo, thumbwheel volume and tone knobs, band switch, telescoping antenna, leather handle, AM, FM, bat **$20.00**

628, horizontal, 1962, charcoal, seven transistors, wedge-shaped case with large right round dial and thumbwheel knob, left perforated grill area, swing handle, AM, bat **$25.00**

629, horizontal, 1963, leather, eight transistors, off-center round dial over large grill area, right tuning and volume knobs, upper left knob, leather handle, AM, bat **$15.00**

Jade

101, vertical, 4¼x2⅝x1⅜", plastic, 10 transistors, upper right front window dial with right side thumbwheel tuning, left side thumbwheel on/off/volume knob, metal perforated grill area, left side braided strap, made in Hong Kong, AM, bat **$10.00**

163, vertical, plastic, six transistors, upper right front window dial with right side thumbwheel tuning, top left thumbwheel on/off/volume knob, lower checkered grill area, AM, bat $10.00

Jaguar

6T-250, vertical, 1960, six transistors, upper right front dial with top thumbwheel tuning, left front volume window with left side thumbwheel knob, lower metal perforated grill area, made in Japan, AM, bat **$40.00**

Jeb

6YR-15A, vertical, plastic, six transistors, upper right see-through dial window with right side thumbwheel tuning, upper left see-through on/off/volume window with left side thumbwheel knob, lower round metal perforated grill area, AM, bat **$30.00**

Jefferson-Travis

JT-D210, vertical, 1961, four transistors, upper right front round dial knob, upper left thumbwheel on/off/volume knob, lower perforated grill area with lower left logo, AM, bat **$30.00**

JT-E212, horizontal, 1961, five transistors, right front window dial with right side tuning, lower right recessed thumbwheel on/off/volume knob, large perforated grill area, AM, bat **$25.00**

JT-F211, vertical, 1961, six transistors, upper front horizontal slide rule dial, large lower perforated grill area with lower left logo, right and left side knobs, AM, bat **$20.00**

JT-G104, horizontal, 1961, seven transistors, upper front horizontal two-band slide rule dial with right side thumbwheel tuning, top left thumbwheel on/off/volume knob, horizontal grill bars, telescoping antenna, AM, SW, bat **$40.00**

JT-G200, vertical, 4½x2¾x1¼", plastic, seven transistors, upper front rectangular two-band window dial with right side thumbwheel tuning, upper left front volume window with left side thumbwheel knob, metal perforated grill area with lower left logo, AM, SW, bat **$25.00**

JT-G204 "Long Distance," vertical, 1961, seven transistors, top right thumbwheel dial knob, top left thumbwheel on/off/volume knob, lower perforated grill area, swing handle, AM, bat **$40.00**

JT-H105, horizontal, 1961, eight transistors, upper front horizontal three-band slide rule dial with thumbwheel tuning, large lower perforated grill area with lower left logo, right side switch, telescoping antenna, AM, 2SW, bat **$30.00**

JT-H204 "Deluxe 8," vertical, 1961, eight transistors, top right thumbwheel dial knob, top left thumbwheel on/off/volume knob, lower perforated grill area, swing handle, AM, bat **$40.00**

Jetstream

JK29B-63A, vertical, 4¼x2½x1¼", plastic, 14 transistors, upper right

front window dial with right side thumbwheel tuning, left side thumbwheel on/off/volume knob, large metal perforated grill area, made in Hong Kong, AM, bat $15.00

Jewel

10, vertical, 1965, plastic, nine transistors, upper left front dial knob, upper right front volume knob, large lower lattice grill area with lower left logo, swing handle, AM, bat **$15.00**

Juliette

AK-8, vertical, 4¼x2⅝x1⅜", plastic, eight transistors, upper right front window dial with right side thumbwheel tuning, top left thumbwheel on/off/volume knob, horizontal grill bars with lower left logo, made in Hong Kong, AM, bat $10.00

APR-306, vertical, 4¼x2⅝x1¼", plastic, solid state, upper right front window dial with right side thumbwheel tuning, top left thumbwheel on/off/volume knob, horizontal grill bars, made in Hong Kong, AM, bat $10.00

AT-65, vertical, 1965, six transistors, upper right front window dial with right side thumbwheel tuning, top left thumbwheel on/off/volume knob, horizontal grill bars, AM, bat **$15.00**

AT-85, vertical, 1965, eight transistors, upper right front window dial with right side thumbwheel tuning, top left thumbwheel on/off/volume knob, horizontal grill bars with lower left logo, AM, bat **$10.00**

AT-105, vertical, 1965, 10 transistors, upper right front window dial with right side thumbwheel tuning, top left thumbwheel on/off/volume knob, horizontal grill bars with lower left logo, AM, bat **$15.00**

DTF-150, horizontal, 1965, 15 transistors, two off-center round dials – one AM, one FM – left perforated grill area with lower left logo, three right knobs, telescoping antenna, handle, AM, FM, bat **$15.00**

SH-516, vertical, 4⅝x2¾x1½", plastic, 16 transistors, upper right front circular window dial with right side thumbwheel tuning, top left thumbwheel on/off/volume knob, recessed circular metal perforated grill area, made in Hong Kong, AM, bat $20.00

Kensington

HT-1268, vertical, 4¼x2½x1¼", plastic, 10 transistors, upper front window dial with right side tuning, top left thumbwheel on/off/volume knob, large metal perforated grill area, AM, bat $20.00

Kent

TR-605, vertical, 1965, six transistors, upper right front window dial with right side thumbwheel tuning, left side thumbwheel on/off/volume knob, large perforated grill area, AM, bat **$15.00**

"Boy's Radio," vertical, 4⅛x2⅝x1¼", plastic, two transistors, top left thumbwheel dial knob, top right thumbwheel on/off/volume knob, large lower metal perforated grill area with lower right logo, left side horizontal telescoping antenna, made in Japan, AM, bat........ **$30.00**

King

"Boy's Radio," vertical, plastic, two transistors, upper right front quarter-round window dial, lower metal perforated grill area, AM, bat $30.00

Knight

KN-2400, horizontal, 1964, nine transistors, top horizontal four-band slide rule dial, pushbuttons and knobs, large front grill area with horizontal bars, two telescoping antennas, handle, AM, FM, 2SW, bat **$20.00**

Kowa

KT-31, vertical, 1960, three transistors, upper right front diamond-shaped window dial with right side thumbwheel tuning, upper left front diamond-shaped volume window with left side thumbwheel knob, lower perforated grill area, AM, bat **$40.00**

KT-62A, horizontal, 1961, six transistors, upper right front window dial with right side thumbwheel tuning, lower right side thumbwheel on/off/volume knob, large left perforated grill area, AM, bat **$30.00**

KT-63, vertical, 4⅛x2⅜x1¼", 1961, plastic, six transistors, upper window dial with upper right dial knob, top left thumbwheel on/off/volume knob, large lower metal perforated grill area, telescoping antenna, swing handle, made in Japan, AM, bat $30.00

KT-66, vertical, 4⅛x2⅝x1¼", 1961, plastic, six transistors, upper window dial with upper right thumbwheel tuning, upper left thumbwheel on/off/volume knob, lower metal perforated grill area, swing handle, made in Japan, AM, bat **$30.00**

KT-67, vertical, 1961, six transistors, upper right front window dial with top right thumbwheel tuning, top left thumbwheel on/off/volume knob, round perforated grill area, AM, bat **$30.00**

KT-91, vertical, 1962, nine transistors, upper window dial with right side thumbwheel tuning, left side thumbwheel on/off/volume knob, lower perforated grill area, AM, bat **$40.00**

KTF-1, horizontal, 7¾x9¼x3¼", 1961, plastic, 11 transistors, top horizontal two-band slide rule dial, lower checkered grill area, AM/FM switch, two telescoping antennas, handle, AM, FM, bat.................................. **$35.00**

KTS-1B, horizontal, 1961, eight transistors, top horizontal two-band slide rule dial with right tuning and left volume knobs, large lower perforated grill area, telescoping antenna, AM, SW, bat **$25.00**

Koyo

KTR834, horizontal, 1965, eight transistors, upper front horizontal two-band slide rule dial with thumbwheel tuning, large lower perforated grill area with lower right logo, telescoping antenna, AM, SW, bat **$20.00**

KTR-1031, horizontal, 1965, 10 transistors, center front vertical slide rule dial, twin speakers with right and left circular perforated grill areas, AM, bat **$25.00**

Kroy

T-1001, vertical, 4½x2⅝x1⅜", plastic, ten transistors, upper right front half-round dial with right side thumbwheel tuning, left side thumbwheel on/off/volume knob, horizontal grill bars, AM, bat $10.00

T-1010, horizontal, 10 transistors, right front vertical two-band slide rule dial with right side thumbwheel tuning, left grill area with horizontal bars, telescoping antenna, AM, FM, bat **$15.00**

Lafayette

17-0101, vertical, 4x2⅝x1¼", 1965, plastic, six transistors, upper right front window dial with right side thumbwheel tuning, left side thumbwheel on/off/volume knob, textured grill area, made in Okinawa, AM, bat $10.00

17-0102, vertical, 1965, 10 transistors, upper front round dial, right thumbwheel knob, lower grill area, AM, bat **$15.00**

17G0104, horizontal, 1965, 15 transistors, upper front horizontal three-band slide rule dial, large lower grill area with lower left logo, four right knobs, two telescoping antennas, handle, AM, FM, SW, bat....... **$15.00**

17G6905L, vertical, 1965, nine transistors, upper right front window dial with right side thumbwheel tuning, large lower grill area with horizontal bars, AM, bat **$10.00**

FS-91, vertical, 4½x2⅞x1⅛", 1961, plastic, nine transistors, upper front quarter-round dial with right side thumbwheel tuning, left side thumbwheel on/off/volume knob, metal perforated grill area, rear fold-out stand, made in Japan, AM, bat $40.00

FS-93, horizontal, 1962, nine transistors, upper front horizontal three-band slide rule dial with right thumbwheel tuning, left thumbwheel on/off/volume knob, upper right band-select window, lower right switches, perforated grill area, telescoping antenna, AM, Weather, Marine, bat................................. **$35.00**

FS-112, vertical, 1959, six transistors, top thumbwheel dial with see-through window, right front thumbwheel on/off/volume knob, lower perforated grill area, AM, bat **$30.00**

FS-129, horizontal, 1963, 10 transistors, right front vertical slide rule dial with lower right side thumbwheel tuning, upper right side thumbwheel on/off/volume knob, perforated grill area, AM, bat **$15.00**

FS-200, horizontal, 1960, six transistors, right front window dial with right side thumbwheel tuning, lower right side thumbwheel on/off/volume knob, perforated grill area, AM, bat **$45.00**

FS-204, vertical, 1961, four transistors, upper left front round window dial with top thumbwheel tuning, top right thumbwheel on/off/volume knob, lower perforated grill area, AM, bat **$30.00**

FS-223, horizontal, 1962, seven transistors, upper front horizontal two-band slide rule dial with thumbwheel tuning, lower left perforated grill area, lower right "star" emblem, telescoping antenna, AM, SW, bat **$35.00**

FS-235, vertical, 1963, six transistors, upper front window dial with thumbwheel tuning, lower perforated grill area, AM, bat **$35.00**

FS-238, horizontal/desk set radio, 1964, seven transistors, left front horizontal dial with two thumbwheel knobs and perforated grill area, two penholders, AM, bat **$25.00**

FS-243, vertical, 1963, six transistors, upper front window dial with thumbwheel tuning, lower round perforated grill area, swing handle, AM, bat **$35.00**

FS-244, horizontal, 1963, eight transistors, upper front horizontal two-band slide rule dial with thumbwheel tuning, large lower perforated grill area, AM, SW, bat **$20.00**

FS-245, horizontal, 1963, 10 transistors, raised center top with horizontal two-band slide rule dial and three pushbuttons, two top knobs, large lower perforated grill area, two telescoping antennas, handle, AM, FM, bat **$35.00**

FS-248, vertical, 1963, six transistors, upper front window dial with thumbwheel tuning, lower perforated grill area with center "L" logo, AM, bat **$15.00**

FS-251, horizontal, 1963, 12 transistors, upper front horizontal three-band slide rule dial, lower perforated grill area, telescoping antenna, handle, AM, FM, SW, bat **$20.00**

FS-252, horizontal, 1963, 12 transistors, upper left front horizontal four-band slide rule dial, four pushbuttons, lower grill area with two knobs, telescoping antenna, handle, AM, 3SW, bat **$20.00**

FS-253, horizontal, 1963, eight transistors, upper front horizontal dial with thumbwheel tuning, center front horizontal cut-outs, lower perforated grill area, AM, bat **$20.00**

FS-258, horizontal, 1964, leather, nine transistors, right front dial knob and left front on/off/volume knob over horizontal grill bars, leather handle, AM, bat **$15.00**

FS-280, horizontal, 1965, 10 transistors, three right front vertical slide rule dial scales with thumbwheel tuning, large left grill area with vertical bars, telescoping antenna, AM, SW, LW, bat **$15.00**

FS-284L, horizontal, 1964, 10 transistors, upper right front horizontal slide rule dial with thumbwheel tuning, left round metal perforated grill area, AM, bat $20.00

FS-305, horizontal, 1965, nine transistors, upper front horizontal two-band slide rule dial with right thumbwheel tuning, left thumbwheel volume knob, right VHF/MW switch, perforated grill area, telescoping antenna, AM, Aviation, bat **$25.00**

TR-1645, vertical, 1965, six transistors, step-back top, upper left front window dial with thumbwheel tuning over horizontal grill bars, right side thumbwheel on/off/volume knob, AM, bat **$10.00**

TR-1660, vertical, 1964, six transistors, upper right front window dial with right side thumbwheel tuning, lower horizontal grill bars, AM, bat **$10.00**

TR-1948, vertical, 1964, nine transistors, upper right front window dial with right side thumbwheel tuning, lower horizontal grill bars, AM, bat **$10.00**

TR-2051, horizontal, 1964, 10 transistors, right front vertical two-band slide rule dial, large left grill area with horizontal bars, telescoping antenna, AM, FM, bat **$15.00**

TR-3047, horizontal, 1964, 10 transistors, three right front vertical slide rule dial scales, large left grill area with vertical bars, telescoping antenna, AM, SW, LW, bat **$15.00**

Lefco

6YR-15A, vertical, 3½x2⅛x⅞", plastic, six transistors, upper right window dial with thumbwheel tuning, upper left on/off/volume window with thumbwheel knob, round metal perforated grill area, AM, bat **$30.00**

Lincoln

24SC054, horizontal, 1965, 10 transistors, right front vertical two-band slide rule dial with thumbwheel tuning, large left grill area with horizontal bars, telescoping antenna, AM, FM, bat **$20.00**

24SC079, vertical, 1965, eight transistors, upper front horizontal slide rule dial with thumbwheel tuning, lower perforated grill area, AM, bat **$20.00**

L640, vertical, 1963, six transistors, upper right front round dial knob, upper left front volume window with top thumbwheel knob, lower perforated grill area, AM, bat **$30.00**

TR-970, vertical, 6x4x2", 1963, plastic, nine transistors, case arches backwards, upper front horizontal three-band slide rule dial with right side thumbwheel tuning, left side thumbwheel volume and tone knobs, lower metal perforated grill area, telescoping antenna, swing handle, SW, LW, MW, bat... $45.00

TR-1055 "Duo Fi," horizontal, 1965, leather, 10 transistors, upper right front window dial, left grill area with cut-outs, H/L switch, leather handle, AM, bat **$10.00**

TR-1844, horizontal, 1964, leather, eight transistors, upper right front window dial, left grill area with circular cut-outs, leather handle, AM, bat **$10.00**

TR-1946, vertical, 1964, nine transistors, upper right front window dial with thumbwheel tuning, large lower grill area with vertical bars, AM, bat **$10.00**

TR-3047, horizontal, 1964, 10 transistors, three right front vertical slide rule dial scales, large left grill area with vertical bars, telescoping antenna, AM, SW, LW, bat **$15.00**

TR-3422, horizontal, 1963, 14 transistors, upper front horizontal three-band slide rule dial, large lower grill area, thumbwheel knobs, three telescoping antennas, handle, AM, FM, SW, bat **$20.00**

TR-4016, horizontal, 1963, ten transistors, upper front horizontal four-band slide rule dial, large lower grill area, thumbwheel knobs, telescoping antenna, handle, AM, 2SW, LW, bat **$20.00**

Linmark

T-40, vertical, 1960, four transistors, upper left front round window dial with top thumbwheel tuning, top right thumbwheel on/off/volume knob, lower perforated grill area, AM, bat **$25.00**

T-61, vertical, 1959, six transistors, left front rectangular window dial with top thumbwheel tuning, top right thumbwheel on/off/volume knob, lower round perforated grill area, AM, bat **$40.00**

T-62, vertical, 1960, six transistors, upper front thumbwheel dial with see-through window, upper right front thumbwheel on/off/volume knob, lower perforated grill area, AM, bat **$35.00**

T-63, vertical, 1960, six transistors, right front window dial with right side thumbwheel tuning, diagonally divided front with perforated grill area, AM, bat **$40.00**

T-80, vertical, 1960, eight transistors, upper right front window dial with right side thumbwheel tuning, upper left front thumbwheel on/off/ volume knob, lower grill area with horizontal slots, swing handle, AM, bat **$35.00**

Lloyd's

6K87B, square, 2½x2½x1¼", plastic, six transistors, upper right side dial knob, lower right side on/off/volume knob, front grill area, left side vinyl strap, AM, bat $40.00

8R-202A "Super Het," horizontal, 1964, eight transistors, upper front horizontal two-band slide rule dial with thumbwheel tuning, large lower perforated grill area with lower right switch, AM, SW, bat **$25.00**

10R-200A3, horizontal, 1964, 10 transistors, lower front horizontal three-band slide rule dial, large upper grill area with horizontal bars, four pushbuttons, telescoping antenna, handle, AM, 2SW, bat **$25.00**

10R-303A, horizontal, 1964, 10 transistors, upper right front horizontal three-band slide rule dial with thumbwheel tuning, left perforated grill area, four top pushbuttons, telescoping antenna, handle, AM, 2SW, bat **$20.00**

108MB "Super Deluxe," horizontal, 1964, 10 transistors, three upper front horizontal slide rule dials, large lower perforated grill area, two speakers, telescoping antenna, AM, 2SW, bat **$25.00**

TF-57L, horizontal, 1964, leather, 11 transistors, two upper left round dials – one AM, one FM – lower perforated grill area with left AM/FM switch, telescoping antenna, leather handle, AM, FM, bat **$15.00**

TF-58, horizontal, 1964, 10 transistors, upper front horizontal two-band slide rule dial with thumbwheel tuning, large lower grill area, telescoping antenna, handle, AM, FM, bat **$20.00**

TF-97, vertical, 1965, nine transistors, the upper front horizontal AM dial forms a right angle in the upper left corner with the left side vertical FM dial, large grill area, telescoping antenna, right side strap, AM, FM, bat **$20.00**

TF-110, vertical, 1964, 10 transistors, upper front horizontal two-band slide rule dial, lower perforated grill area, telescoping antenna, handle, AM, FM, bat **$15.00**

TF-129L, horizontal, 1964, 12 transistors, three upper front horizontal slide rule dials, large lower perforated grill area, top pushbuttons, telescoping antenna, handle, AM, FM, SW, bat **$20.00**

TF-311, horizontal, 1965, 10 transistors, two upper front round dials and three knobs, lower grill area with horizontal bars and center logo, telescoping antenna, handle, AM, FM, SW, bat **$20.00**

TF-911, horizontal, 1964, nine transistors, upper right front round two-band dial with top right thumbwheel tuning, off-center top thumbwheel on/off/volume knob, lower AM/FM switch, left grill area, telescoping antenna, AM, FM, bat **$20.00**

TF-912, horizontal, 3¾x6⅜x1¾", 1965, plastic, nine transistors, upper right front round dial with top right thumbwheel tuning, off-center top thumbwheel on/off/volume knob, lower AM/FM switch, left grill area with horizontal bars, telescoping antenna, right side vinyl strap, made in Japan, AM, FM, bat $15.00

TF-990L, horizontal, 1965, leather, nine transistors, off-center vertical two-band slide rule dial, two right knobs and AM/FM switch, left perforated grill area, telescoping antenna, handle, AM, FM, bat **$15.00**

TR-6KA, vertical, 1964, six transistors, upper front round window dial with right side thumbwheel tuning, lower grill area with vertical slots, AM, bat **$15.00**

TR-6KB, horizontal, 1965, six transistors, upper right front window dial with right side thumbwheel tuning, lower right side thumbwheel on/off/volume knob, horizontal grill bars, AM, bat **$10.00**

TR-6L, vertical, 1964, six transistors, upper left front window dial with thumbwheel tuning, upper right front thumbwheel on/off/volume knob, lower perforated grill area, AM, bat **$20.00**

TR-6P, vertical, 1965, six transistors, upper right front round window dial with thumbwheel tuning, large lower grill area with horizontal bars, right side strap, AM, bat **$10.00**

TR-6T, vertical, 1964, six transistors, upper right front round window dial with thumbwheel tuning, large lower grill area with horizontal bars, right side strap, AM, bat **$10.00**

TR-8KA, vertical, 1964, eight transistors, upper right front window dial with right side thumbwheel tuning, upper left front on/off/volume window with left side thumbwheel knob, large grill area with horizontal slots, AM, bat **$15.00**

TR-8KB, horizontal, 1965, eight transistors, upper right front window dial with right side thumbwheel tuning, lower right side thumbwheel on/off/volume knob, horizontal grill bars, AM, bat **$10.00**

TR-8L, vertical, 1964, eight transistors, upper left window dial with top left thumbwheel tuning, upper right thumbwheel on/off/volume knob, lower perforated grill area, AM, bat **$20.00**

TR-10K, horizontal, 1965, 10 transistors, upper right front window dial with right side thumbwheel tuning, lower right side thumbwheel on/off/volume knob, horizontal grill bars, top right strap, AM, bat **$10.00**

TR-10L, horizontal, 1964, leather, 10 transistors, upper left round dial, two right knobs, large lower grill area with horizontal bars and center logo, leather handle, AM, bat **$10.00**

TR-10N, vertical, 1965, 10 transistors, upper right front window dial, left knob, large lower grill area with horizontal bars, AM, bat **$10.00**

TR-12L, horizontal, 12 transistors, upper right front round window dial with right side thumbwheel tuning, right side thumbwheel on/off/volume knob, front grill area with horizontal bars, AM, bat $15.00

TR-86L, horizontal, 1965, leather, eight transistors, right front thumbwheel dial, top left thumb-wheel on/off/volume knob, horizontal grill bars, handle, AM, bat **$15.00**

TR-89L, horizontal, 1964, leather, eight transistors, right front thumbwheel dial, upper left on/off/volume knob, large grill area with horizontal slots, leather handle, AM, bat **$15.00**

TRA-10, vertical, 1965, 10 transistors, upper right front round dial with right side thumbwheel tuning, lower grill area, top rectangular handle, AM, bat **$20.00**

"Boy's Radio," vertical, 4⅛x2½x1¼", plastic, two transistors, upper right front window dial with right side thumbwheel tuning, left side thumbwheel on/off/volume knob, lower metal perforated grill area, made in Japan, AM, bat **$25.00**

Maco

AB-100, vertical, 1960, six transistors, right window dial with right side thumbwheel tuning, upper right front thumbwheel on/off/volume knob, lower perforated grill area with lower left logo, AM, bat **$25.00**

AB-175M, horizontal, 1962, eight transistors, upper front horizontal two-band slide rule dial with right thumbwheel tuning, top left thumbwheel on/off/volume knob, large perforated grill area with lower left logo, telescoping antenna, AM, SW, bat $25.00

T-16, horizontal, 1960, six transistors, upper right front dial with right side thumbwheel tuning, lower right front on/off/volume window with right side thumbwheel knob, left perforated grill area with lower left logo, AM, bat $40.00

Magnavox

2AM-70, vertical, 1964, seven transistors, upper off-center window dial with right side thumbwheel tuning, right side thumbwheel on/off/volume knob, perforated grill area with lower left logo, AM, bat $30.00

2-AM-80, vertical, 4⅛x2½x1", 1963, plastic, eight transistors, upper front window dial with right side thumbwheel tuning, right side thumbwheel on/off/volume knob, large metal perforated grill area with lower right logo, made in Japan, AM, bat $25.00

2AM081, vertical, 4⅜x2⅝x1½", plastic, eight transistors, upper front horizontal slide rule dial with right side thumbwheel tuning, right side thumbwheel on/off/volume knob, metal perforated grill area with lower right logo, made in Japan, AM, bat $20.00

2-AM-802, horizontal, 3x4¾x1½", plastic, eight transistors, upper front horizontal slide rule dial, large lower metal perforated grill area with right thumbwheel tuning and volume knobs, strap, AM, bat $20.00

2-AM-811, vertical, eight transistors, upper right front dial knob, left side thumbwheel on/off/volume knob, left vertical grill area with vertical bars, right side strap, AM, bat $15.00

AM-2, horizontal, 1957, Magnavox's first transistor radio, large right front round dial, lower right side thumbwheel on/off/volume knob, large perforated grill area, AM, bat **$125.00**

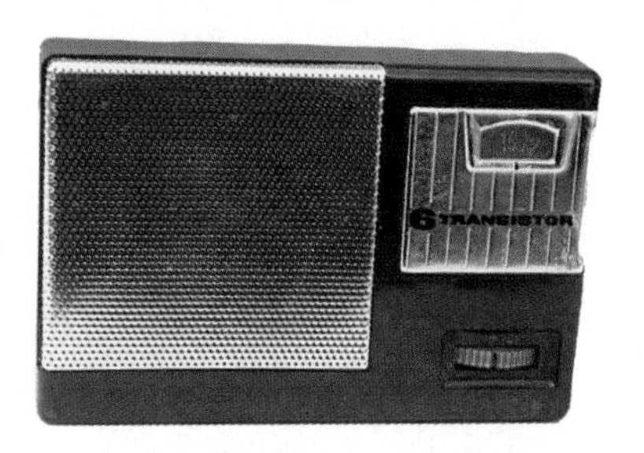

AM-22, horizontal, 2¾x4¼x1", 1960, plastic, six transistors, upper right front window dial with right side thumbwheel tuning, lower right front thumbwheel on/off/volume knob, left metal perforated grill area, AM, bat $35.00

AM-23, vertical, 4⅛x2⅝x1¼", 1960, plastic, six transistors, upper left round dial knob, right side thumbwheel on/off/volume knob, lower metal perforated grill area, telescoping antenna, made in Japan, AM, bat $35.00

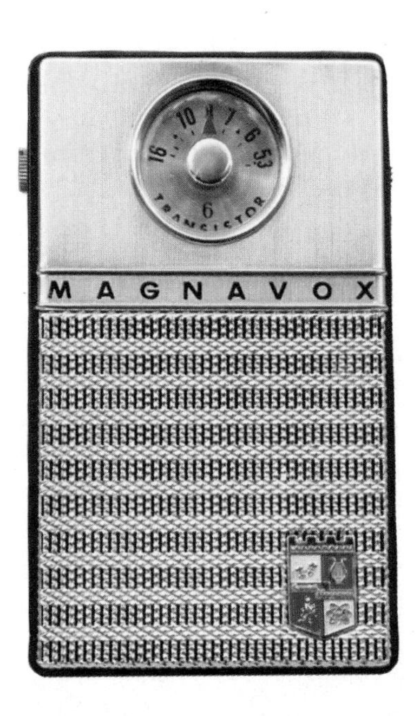

AM-60, vertical, 4⅛x2⅝x1", 1961, plastic, six transistors, upper round dial with right side thumbwheel tuning, right side thumbwheel on/off/ volume knob, metal perforated grill area, rear fold-out stand, made in Japan, AM, bat $25.00

AM-61, vertical, 1965, plastic, six transistors, upper front horizontal slide rule dial with right side thumbwheel tuning, right side thumbwheel on/ off/volume knob, lower metal perforated grill area with lower right logo, AM, bat **$20.00**

AM-62, vertical, 1963, upper front dial with right side thumbwheel tuning, right side thumbwheel on/ off/volume knob, metal perforated grill area with lower left logo, AM, bat **$25.00**

AM-64, horizontal, 3¼x6x1¾", 1963, plastic, six transistors, right front vertical slide rule dial with right side thumbwheel tuning, right side thumbwheel on/off/ volume knob, large perforated grill area, AM, bat **$20.00**

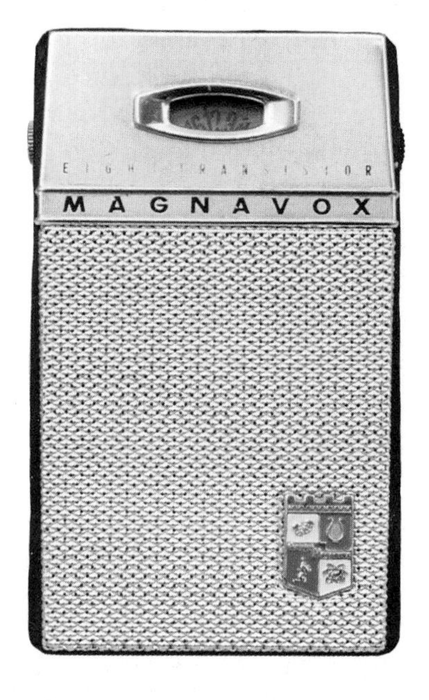

AM-80, vertical, 4⅛x2½x1", 1961, plastic, eight transistors, upper front window dial with right side thumbwheel tuning, right side thumbwheel on/off/volume knob, large metal perforated grill area with lower right logo, rear fold-out stand, AM, bat $25.00

AM-81, vertical, 4⅜x2⅝x1½", plastic, eight transistors, upper front horizontal slide rule dial with right side thumbwheel tuning, right side thumbwheel on/off/volume knob,

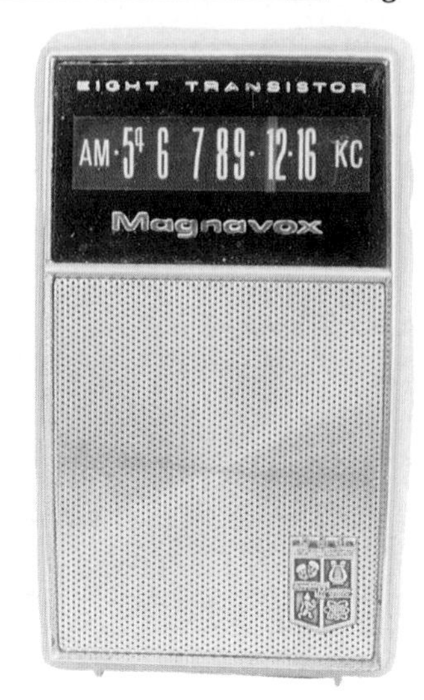

metal perforated grill area with lower right logo, made in Japan, AM, bat $20.00

AM-82 "Envoy," horizontal, 1964, leather, eight transistors, off-center vertical slide rule dial, two right knobs, left perforated grill area with lower left logo, handle, AM, bat **$15.00**

AM-83 "Safari," horizontal, 6x8x 3¾", 1964, leather, eight transistors, upper front horizontal slide rule dial, two right knobs, lower perforated grill area with lower left logo, handle, AM, bat **$15.00**

AM-85, horizontal, 4¼x7¼x1¾", 1963, plastic, eight transistors, right front dial with thumbwheel tuning, right thumbwheel on/off/volume knob, left perforated grill area with left logo, AM, bat **$25.00**

AT-61, horizontal, 5⅞x10⅝x4⅝", 1961, available in black, light green, or sand plastic, six transistors, upper right front horizontal slide rule dial, large lower grill area with logo and two knobs, AM, bat **$20.00**

AW-24, horizontal, 1960, seven transistors, upper front horizontal two-band slide rule dial with top thumbwheel tuning, top thumbwheel on/off/volume knob, metal perforated grill area with lower right logo, telescoping antenna, AM, SW, bat **$25.00**

AW-88 "Constellation," horizontal, 1964, leather, eight transistors, upper front horizontal three-band slide rule dial, four right knobs, left metal perforated grill area with lower left logo, top knob, handle, AM, SW, LW, bat **$25.00**

FM-90, horizontal, 1962, 10 transistors, upper front horizontal two-band slide rule dial with upper right front thumbwheel tuning, upper left front thumbwheel on/off/volume knob, top pushbuttons, perforated grill area with lower left logo, two telescoping antennas, handle, AM, FM, bat **$25.00**

FM-91, horizontal, 8½x10¼x3¾", 1965, 10 transistors, upper left front horizontal two-band slide rule dial, two upper right knobs and one switch, lower grill area with lower left logo, telescoping antenna, handle, AM, FM, bat **$20.00**

FM-92, vertical, 1965, nine transistors, two upper front window dials – right AM, left FM – large perforated grill area with lower right logo, telescoping antenna, AM, FM, bat...... **$15.00**

FM-95 "Constellation," horizontal, 1963, nine transistors, upper front horizontal three-band slide rule dial, four right knobs, large grill area with lower left logo, two telescoping antennas, handle, AM, FM, SW, bat **$25.00**

FM-97 "Celestial," horizontal, 1963, nine transistors, upper front horizontal four-band slide rule dial, four right knobs, large grill area with lower left logo, telescoping antenna, handle, AM, FM, 2SW, bat..... **$25.00**

Majestic

6G780, vertical, plastic, six transistors, upper left front round dial, right side thumbwheel on/off/volume knob, metal perforated grill area with lower right bird logo, AM, bat **$40.00**

FX-408, horizontal, plastic, 11 transistors, upper front horizontal two-band slide rule dial with upper right thumbwheel tuning, left thumbwheel on/off/volume knob, right FM/AM switch, lower metal perforated grill area with lower right eagle logo, AM, FM, bat $25.00

"Super Eighty," vertical, 5¾x3½x 1½", plastic, upper right front dial knob, upper left front on/off/volume knob, lower horizontal grill bars, swing handle, made in USA, AM, bat $25.00

Mantola

M4D, vertical, 5x3x1¼", plastic, four transistors, similar to the Regency TR-1, large upper right front round brass dial knob with orbiting electrons, upper left thumbwheel on/off/volume knob, lower perforated grill area, AM, bat **$225.00**

Marvel

6YR-05, vertical, 1961, six transistors, upper right front round dial with right side thumbwheel tuning, upper left front on/off/volume window with left side thumbwheel knob, perforated grill area, AM, bat **$30.00**

8YR-10A, vertical, 1962, eight transistors, upper front horizontal slide rule dial with right side thumbwheel tuning, right side thumbwheel on/off/volume knob, lower round perforated grill area, AM, bat **$35.00**

Mascot

RE-60, vertical, 1965, six transistors, upper right front window dial with right side thumbwheel tuning, lower vertical grill bars, right side strap, AM, bat **$10.00**

TR2, vertical, plastic, upper right front thumbwheel dial knob, upper left front thumbwheel on/off/volume knob, lower metal perforated grill area, made in Japan, AM, bat **$25.00**

Master-Craft

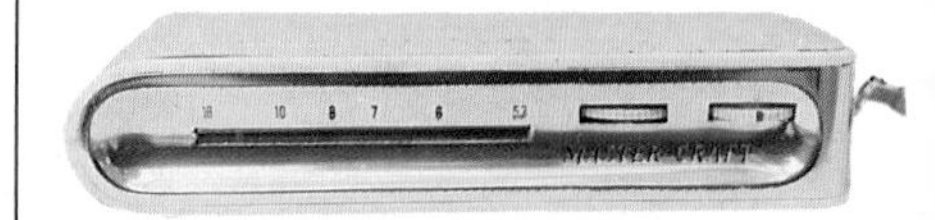

TF-810, horizontal, 1⅜x6½x2⅜", leather/plastic/metal, eight transistors, concave front, left horizontal slide rule dial with thumbwheel tuning, thumbwheel on/off/volume knob, right side braided strap, made in Hong Kong, AM, bat $20.00

Masterwork

M2100TR "Galaxy III," horizontal, 1963, nine transistors, upper front horizontal three-band slide rule dial, top pushbuttons and thumbwheel knobs, lower checkered grill area, telescoping antenna, handle, AM, FM, SW, bat **$20.00**

M2810, horizontal, 1964, 10 transistors, upper front horizontal two-band slide rule dial, top left thumbwheel on/off/volume knob, large lower perforated grill area, telescoping antenna, AM, FM, bat **$15.00**

M2812, horizontal, 1965, leather, eight transistors, upper left front horizontal slide rule dial, two right knobs, large lower perforated grill area, leather handle, AM, bat **$10.00**

M2815, horizontal, 1964, 10 transistors, two right front round dials – one AM, one FM – left perforated grill area, telescoping antenna, handle, AM, FM, bat **$20.00**

M3005, horizontal/clock radio, 1964, six transistors, right front window dial with top thumbwheel tuning, left alarm clock face, AM, bat **$10.00**

Matsushita

DT-495, horizontal, 1962, six transistors, upper right front thumbwheel dial knob, lower right on/off/volume knob, left circular grill area with horizontal bars and lower left logo, AM, bat **$30.00**

T-7, vertical, 3¾x2½x1⅛", plastic, seven transistors, upper front horizontal slide rule dial with right side thumbwheel tuning, right side thumbwheel on/off/volume knob, metal grill area with horizontal slots, rear fold-out stand, made in Japan, AM, bat $30.00

T-13, vertical, 1961, six transistors, upper front horizontal slide rule dial with right side thumbwheel tuning, upper left thumbwheel on/off/volume knob, lower perforated grill area with lower right logo, AM, bat **$25.00**

T-22M, horizontal, 1962, eight transistors, upper front horizontal two-band slide rule dial with thumbwheel tuning, perforated grill area with MW/SW switch and battery window, telescoping antenna, AM, SW, bat **$30.00**

T-22U, horizontal, 1962, eight transistors, upper front horizontal two-band slide rule dial with thumbwheel tuning, perforated grill area with MW/SW switch and battery window, telescoping antenna, AM, SW, bat **$30.00**

T-30, horizontal, 6⅝x10x2⅞", 1961, plastic, nine transistors, upper front horizontal two-band slide rule dial with thumbwheel tuning, upper left AM/FM switch, lower perforated grill area, telescoping antenna, handle, AM, FM, bat **$25.00**

T-41M, horizontal, 1962, eight transistors, top horizontal two-band slide rule dial with top right thumbwheel tuning, lower right side thumbwheel on/off/volume knob, large perforated grill area with lower right logo, telescoping antenna, AM, SW, bat **$25.00**

T-41U, horizontal, 1962, eight transistors, top horizontal two-band slide rule dial with top right thumbwheel tuning, lower right side thumbwheel on/off/volume knob, large perforated grill area with lower right logo, telescoping antenna, AM, SW, bat $25.00

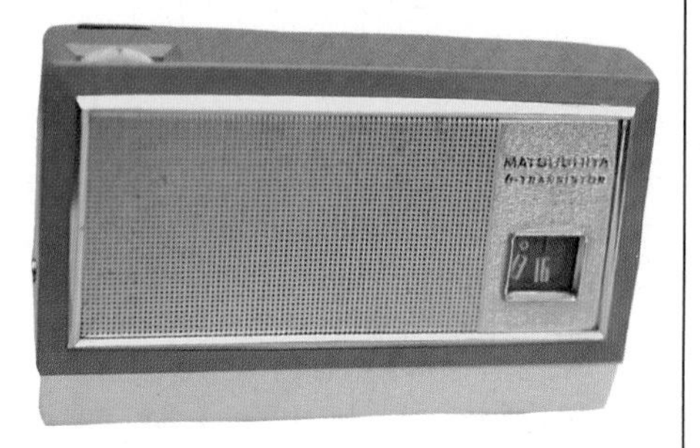

T-50, horizontal, 3¼x5¾x1½", 1962, plastic, six transistors, right front window dial with right side thumbwheel tuning, top left thumbwheel on/off/volume knob, large metal perforated grill area, AM, bat $30.00

T-66, horizontal, 1962, eight transistors, upper front horizontal three-band slide rule dial, large lower perforated grill area, right side switch, telescoping antenna, handle, AM, 2SW, bat **$25.00**

T-70M, horizontal, 1962, eight transistors, right front horizontal two-band slide rule dial with thumbwheel tuning, thumbwheel on/off/volume knob, left grill area with horizontal bars and upper left logo, telescoping antenna, handle, AM, SW, bat **$25.00**

T-70U, horizontal, 1962, eight transistors, right front horizontal two-band slide rule dial with thumbwheel tuning, thumbwheel on/off/volume knob, left grill area with horizontal bars and upper left logo, telescoping antenna, handle, AM, SW, bat **$25.00**

T-92 "Portalarm," vertical/watch radio, 1962, six transistors, upper right side thumbwheel dial knob with wedge-shaped indent, upper left seven-jewel alarm watch face, top left watch stem, top right switch, perforated grill area with lower left logo, AM, bat **$100.00**

Mellow-Tone

"Boy's Radio," vertical, 4x2½x1¼", plastic, two transistors, rounded upper right corner, upper right front dial knob, lower metal perforated grill area, AM, bat **$45.00**

Melodic

GT-586, vertical, 1961, six transistors, upper left front round dial knob, upper right front thumbwheel on/off/volume knob, lower perforated grill area, AM, bat **$25.00**

MT-69, vertical, 1961, six transistors, upper right front window dial with right side thumbwheel tuning, perforated grill area, AM, bat **$15.00**

Metz

Babyphon 102, horizontal/radio-phono, plastic, left front round dial knob, right front on/off/volume knob, horizontal grill bars, lift top, inner phono, handle, made in West Germany **$35.00**

Midland

10-310, horizontal, 1964, 10 transistors, upper right front window dial with right side thumbwheel tuning, lower right side thumbwheel on/off/volume knob, large perforated grill area, AM, bat **$20.00**

10-408, horizontal, 1964, eight transistors, right front round dial knob, lower right on/off/volume knob, large left perforated grill area, handle, AM, bat **$10.00**

10-410, horizontal, 1964, leather, 10 transistors, right front round dial, lower right tone switch, upper left on/off/volume knob, large perforated grill area, leather handle, AM, bat **$10.00**

10-440, horizontal, 1964, 12 transistors, upper front horizontal two-band slide rule dial, two knobs, large lower perforated grill area, telescoping antenna, handle, AM, FM, bat **$15.00**

11-406, horizontal, 1964, six transistors, right front round dial with thumbwheel tuning, thumbwheel on/off/volume knob, left grill area, AM, bat **$15.00**

Million

G-601, (top right) horizontal, 3x5¼ x1½", plastic, six transistors, "mother-of-pearl" right front with thumbwheel dial knob, top thumbwheel on/off/volume knob, left metal perforated grill area, AM, bat $45.00

Minute Man

6T-170, vertical, 1960, six transistors, top dial and on/off/volume knobs, two front perforated semi-circular wrap-around grill areas, AM, bat **$65.00**

Mitchell

1101, horizontal, 5x3x1½", 1956, suntan leather, four transistors, upper right front round dial knob and upper left thumbwheel on/off/volume knob over large perforated grill area, hinged back, AM, bat ... $250.00 (Photo courtesy of Larry Mitchell)

1102, horizontal, 5x3x1½", 1956, simulated alligator leather, four transistors, upper right front round dial knob and upper left thumbwheel on/off/volume knob over large perforated grill area, hinged back, AM, bat **$250.00**

1103, horizontal, 5x3x1½", 1956, antique white leather, four transistors, upper right front round dial knob and upper left thumbwheel on/off/volume knob over large perforated grill area, hinged back, AM, bat **$250.00**

Mitsubishi

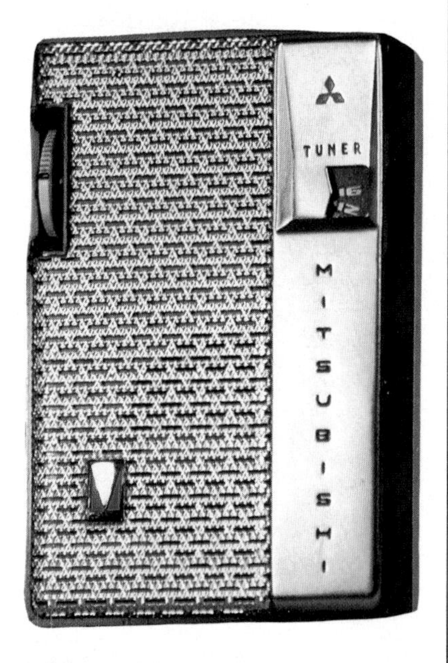

6X-720, vertical, plastic, upper right front recessed window dial with thumbwheel tuning, upper left front thumbwheel on/off/volume knob, large vertical metal textured and perforated grill area with lower left logo, AM, bat $35.00

7X-164, vertical, 1965, seven transistors, upper right front round window dial with right side thumbwheel tuning, lower perforated grill area with lower left logo, AM, bat **$15.00**

9X-628, horizontal, 1965, leather, nine transistors, right front vertical slide rule dial with thumbwheel tuning, upper left front knob, horizontal grill bars, leather handle, AM, bat................. **$15.00**

9X-980, horizontal, 1965, nine transistors, right front round dial with right side thumbwheel tuning, left perforated grill area with lower left logo, AM, bat **$15.00**

FX-233 "Three Diamonds," horizontal, 1965, 13 transistors, upper front horizontal three-band slide rule dial, four thumbwheel knobs, MW/SW/FM switch, battery window, perforated grill area, telescoping antenna, AM, FM, SW, bat **$20.00**

FX-412, horizontal, 1965, nine transistors, upper front horizontal two-band slide rule dial with right side thumbwheel tuning, right side thumbwheel on/off/volume knob, large perforated grill area with upper left logo, telescoping antenna, AM, FM, bat **$15.00**

MMA

6TP-317 "High Fidelity," vertical, 1963, six transistors, upper left front window dial with right side thumbwheel tuning, right side thumbwheel on/off/volume knob, front perforated grill area, AM, bat **$15.00**

8TP-412 "High Sensitivity," vertical, 1963, eight transistors, upper front window dial with right side thumbwheel tuning, right side thumbwheel on/off/volume knob, front perforated grill area with center logo, swing handle, AM, bat **$25.00**

8TP-416, vertical, 1963, six transistors, upper front window dial with right side thumbwheel tuning, right side thumbwheel on/off/volume knob, front perforated grill area, AM, bat **$15.00**

8TP-802M, horizontal, 1962, nine transistors, upper front horizontal three-band slide rule dial with right thumbwheel tuning, two left thumbwheel knobs – one volume, one tone – right side switch, perforated grill area, telescoping antenna, AM, Marine, SW, bat **$30.00**

8TP-905, horizontal, 1963, eight transistors, right front thumbwheel dial, upper left thumbwheel on/off/volume knob, handle, AM, bat **$20.00**

602, vertical, 1963, six transistors, upper right thumbwheel dial, top left thumbwheel on/off/volume knob, perforated grill area, AM, bat **$20.00**

F100, horizontal, 1963, 11 transistors, upper front horizontal three-band slide rule dial, two right side knobs, two left front thumbwheel knobs, horizontal grill bars, two telescoping antennas, handle, AM, FM, SW, bat **$20.00**

F-140, horizontal, 1963, 10 transistors, right front thumbwheel dial, upper right AM/FM switch, upper left thumbwheel on/off/volume knob, perforated grill area, telescoping antenna, strap, AM, FM, bat **$20.00**

TF-52, horizontal, 1963, 11 transistors, upper front horizontal two-band slide rule dial, top knobs, large lower grill area, two telescoping antennas, handle, AM, FM, bat **$25.00**

Monacor

RE-3B "Deluxe," horizontal, 1964, eight transistors, upper front horizontal three-band slide rule dial with right thumbwheel tuning, large lower perforated grill area, telescoping antenna, AM, 2SW, bat **$30.00**

RE-606, vertical, 1964, six transistors, upper right front window dial with thumbwheel tuning, lower vertical grill bars, right side strap, AM, bat **$15.00**

RE-612, vertical, 1963, six transistors, upper left front window dial with right side thumbwheel tuning, right side thumbwheel on/off/volume knob, large perforated circular grill area, AM, bat **$25.00**

RE-613, vertical, 1963, seven transistors, upper front window dial with right side thumbwheel tuning, right side thumbwheel on/off/volume knob, perforated grill area, swing handle, AM, bat **$25.00**

RE-808, vertical, 1964, eight transistors, upper right front window dial with right side thumbwheel tuning, upper left front on/off/volume window with left side thumbwheel knob, perforated grill area, AM, bat **$20.00**

RE-1010 "Sportsman," horizontal, 1964, leather, 10 transistors, right front thumbwheel dial, top left on/off knob, large perforated grill area, handle, AM, bat **$15.00**

RE-1200, horizontal, 1965, nine transistors, right front two-band thumbwheel dial, lower right thumbwheel on/off/volume knob, horizontal grill bars, telescoping antenna, AM, FM, bat **$15.00**

RE-1250L, horizontal, 1965, leather, ten transistors, right front two-band thumbwheel dial, upper right FM/AM switch, upper left thumbwheel on/off/volume knob, perforated grill area, telescoping antenna, handle, AM, FM, bat **$15.00**

RE-1700, horizontal, 1964, 11 transistors, upper front horizontal two-band slide rule dial, large lower grill area with three knobs, telescoping antenna, handle, AM, FM, bat ... **$15.00**

Monarch

90, vertical, 4⅝x3x1¼", plastic, nine transistors, upper front window dial with right side thumbwheel tuning, left side thumbwheel on/off/volume knob, large perforated grill area with lower right logo, rear fold-out stand, made in Japan, AM, bat **$35.00**

RE-760, horizontal, 1964, seven transistors, right front round dial knob, right side knob, feet, handle, AM, bat **$15.00**

RE-1050, horizontal, 1964, 10 transistors, lower right front dial, upper right on/off/volume knob, large left perforated grill area, handle, AM, bat **$15.00**

Motorola

6X28B, horizontal, 1959, blue, six transistors, "jet plane" design molded into front panel over vertical grill bars, right window dial with right side thumbwheel tuning, right side thumbwheel on/off/volume knob, AM, bat **$50.00**

6X28N, horizontal, 1959, mocha, six transistors, "jet plane" design molded into front panel over vertical grill bars, right window dial with right side thumbwheel tuning, right side thumbwheel on/off/volume knob, AM, bat **$50.00**

6X28P, horizontal, 1959, pink, six transistors, "jet plane" design molded into front panel over vertical grill bars, right window dial with right side thumbwheel tuning, right side thumbwheel on/off/volume knob, AM, bat **$50.00**

6X28W, horizontal, 1959, antique white, six transistors, "jet plane" design molded into front panel over vertical grill bars, right window dial with right side thumbwheel tuning, right side thumbwheel on/off/volume knob, AM, bat **$50.00**

6X31C, horizontal, 4x6$\frac{3}{8}$x1$\frac{7}{8}$", 1957, metal case, right front dial knob, upper right on/off/volume knob, plastic grill area with vertical bars, large swing handle, AM, bat $50.00

6X31N, horizontal, 1957, metal case, right front dial knob, upper right on/off/volume knob, plastic grill area with vertical bars, large swing handle, AM, bat **$50.00**

6X31R, horizontal, 4x6$\frac{3}{8}$x1$\frac{7}{8}$", 1957, metal case, right front dial knob, upper right on/off/volume knob, plastic grill area with vertical bars, large swing handle, AM, bat **$50.00**

6X32E-1, horizontal, 4$\frac{1}{8}$x6$\frac{1}{2}$x1$\frac{3}{4}$", metal case, right front dial knob, upper right on/off/volume knob, metal perforated grill area with lower left logo, large plastic swing handle, AM, bat **$50.00**

6X39A-2 "Weatherama," horizontal, 1958, six transistors, right front two-band dial knob, upper right on/off/volume knob, left perforated grill area, large swing handle, AM, Beacon, bat **$80.00**

7X23E "Power 10," horizontal, 4x6$\frac{1}{2}$x2", 1959, navy blue, seven transistors, metal "jet plane" design molded into front panel over perforated grill area, right window dial with right side thumbwheel tuning, top right on/off/volume knob, large swing handle, AM, bat $110.00

7X24S, horizontal, 1959, suntan, seven transistors, metal "jet plane" design molded into front panel over perforated grill area, right window dial with right side thumbwheel tuning, top right on/off/volume knob, large swing handle, AM, bat **$110.00**

7X24W, horizontal, 1959, antique white, seven transistors, metal "jet plane" design molded into front panel over perforated grill area, right window dial with right side thumbwheel tuning, top right on/off/volume knob, back panel has embossed stars, large swing handle, AM, bat **$150.00**

7X25P, vertical, 1959, salmon, seven transistors, upper front round dial knob over horizontal grill bars, lower left on/off/volume knob, handle, AM, bat **$30.00**

7X25W, vertical, 1959, antique white, seven transistors, upper front round dial knob over horizontal grill bars, lower left on/off/volume knob, handle, AM, bat **$30.00**

8X26E, vertical, 7x4½x2½", charcoal plastic, eight transistors, upper front round dial knob, lower left on/off/volume knob, horizontal grill bars with logo, swing handle, AM, bat **$30.00**

8X26S, vertical, 7x4½x2½", maple sugar plastic, eight transistors, upper front round dial knob, lower left on/off/volume knob, horizontal grill bars with logo, swing handle, AM, bat $30.00

56T1, horizontal, 3½x5½x1½", 1956, five transistors, Motorola's first transistor radio, lower right front round dial knob, upper right on/off/volume knob, perforated grill area with lower left logo, large swing handle with built-in antenna, AM, bat **$150.00**

66T1, horizontal, 1958, six transistors, lower right front round dial knob, upper right on/off/volume knob, perforated grill area with lower left logo, large swing handle, AM, bat **$95.00**

76T1, horizontal, 1957, charcoal leatherette, seven transistors, upper right front dial knob, upper left front on/off/volume knob, large metal perforated grill area, rotatable handle, AM, bat **$40.00**

76T2, horizontal, 1957, brown leatherette, seven transistors, upper right front dial knob, upper left front on/off/volume knob, large metal perforated grill area, rotatable handle, AM, bat $40.00

AX4B, horizontal, 1961, light blue front/white back, six transistors, large right front round dial and two knobs over lattice grill area, left logo, AM, bat **$20.00**

AX4G, horizontal, 1961, willow green front/antique white back, six transistors, large right front round dial and two knobs over lattice grill area, left logo, AM, bat **$20.00**

AX4N, horizontal, 1961, sagebrush brown front/antique white back, six transistors, large right front round dial and two knobs over lattice grill area, left logo, AM, bat **$20.00**

CD-XT18B, horizontal, 6⅜x10¼x3⅛", 1962, blue, equipped with Civil Defense antenna terminal to allow emergency reception in fallout shelters, large right front dial knob over large lattice grill area, left on/off/volume knob, pop-up handle, AM, bat **$20.00**

CD-XT18S, horizontal, 6⅜x10¼x3⅛", 1962, tan, equipped with Civil Defense antenna terminal to allow emergency reception in fallout shelters, large right front dial knob over large lattice grill area, left on/off/volume knob, pop-up handle, AM, bat **$20.00**

CX1B, horizontal/clock radio, 1961, blue, six transistors, fold-down front, inner right window dial and two knobs, left clock face, center grill area, AM, bat **$25.00**

CX1E, horizontal/clock radio, 1961, black, six transistors, fold-down front, inner right window dial and two knobs, left clock face, center grill area, AM, bat......................... **$25.00**

CX1W, horizontal/clock radio, 1961, white, six transistors, fold-down front, inner right window dial and two knobs, left clock face, center grill area, AM, bat.......................... **$25.00**

CX2N "Tandem," horizontal/clock radio with removable portable radio unit, 1963, plastic, six transistors, radio unit has upper left front half-round dial with right side thumbwheel tuning, right side thumbwheel on/off/volume knob, large round metal perforated grill area, AM, bat.
radio without clock $25.00
radio with clock $50.00

L12G "Power 8," horizontal, 1960, plastic, six transistors, right front round dial, left on/off/volume knob, lower grill area with horizontal bars and lower left logo, handle, AM, bat $20.00

L12N "Power 8," horizontal, 1960, plastic, six transistors, right front round dial, left on/off/volume knob, lower grill area with horizontal bars and lower left logo, handle, AM, bat **$20.00**

L13S, horizontal, 1960, seven transistors, right front round dial, left on/off/volume knob, lower grill area with horizontal bars and lower left logo, handle, AM, bat **$20.00**

L14E, horizontal, 1960, eight transistors, right front round dial, left on/off/volume knob, lower grill area with horizontal bars and lower left logo, handle, AM, bat **$20.00**

X11B, vertical, 4x2¼x1", 1960, six transistors, upper front oval window dial with right side thumbwheel tuning, left side thumbwheel on/off/volume knob, perforated grill area with "M" logo, rear fold-out stand, AM, bat **$35.00**

X11E, vertical, 4x2¼x1", 1960, plastic, six transistors, upper front oval window dial with right side thumbwheel tuning, left side thumbwheel on/off/volume knob, perforated grill area with "M" logo, rear fold-out stand, AM, bat $35.00

X11R, vertical, 4x2¼x1", 1960, six transistors, upper front oval window dial with right side thumbwheel tuning, left side thumbwheel on/off/volume knob, perforated grill area with "M" logo, rear fold-out stand, AM, bat **$35.00**

X12A "Power Eight," horizontal, 1960, plastic, six transistors, lower right front round window dial with lower right side thumbwheel tuning, upper right side thumbwheel on/off/volume knob, left vertical grill bars, AM, bat **$30.00**

X12A-1 "Power Eight," horizontal, 3½x6x1⅞", 1960, gray plastic, six transistors, lower right front round window dial with lower right side thumbwheel tuning, upper right side thumbwheel on/off/volume knob, left vertical grill bars, AM, bat **$30.00**

X12E-1 "Power Eight," horizontal, 3½x6x1⅞", 1960, smoke plastic, six transistors, lower right front round window dial with lower right side thumbwheel tuning, upper right side thumbwheel on/off/volume knob, left vertical grill bars, AM, bat **$30.00**

X14B, vertical, 4x2¾x1¼", 1960, blue plastic, six transistors, upper left front window dial with thumbwheel tuning, lower oversized round metal perforated grill area, rear fold-out stand, AM, bat **$50.00**

X14E, vertical, 4x2¾x1¼", 1960, black plastic, six transistors, upper left front window dial with thumbwheel tuning, lower oversized round metal perforated grill area, rear fold-out stand, AM, bat $50.00

X14R, vertical, 4x2¾x1¼", 1960, red plastic, six transistors, upper left front window dial with thumbwheel tuning, lower oversized round metal perforated grill area, rear fold-out stand, AM, bat **$50.00**

X14W, vertical, 4x2¾x1¼", 1960, white plastic, six transistors, upper left front window dial with thumbwheel tuning, lower oversized round metal perforated grill area, rear fold-out stand, AM, bat **$50.00**

X15A, vertical, 4x2¾x1½", 1961, gray/blue plastic, six transistors, upper left front window dial with thumbwheel tuning, lower metal perforated grill area, rear swing-out stand, AM, bat **$50.00**

X15E, vertical, 4x2¾x1½", 1961, black plastic, six transistors, upper left front window dial with thumbwheel tuning, lower metal perforated grill area, rear swing-out stand, AM, bat **$50.00**

X15N, vertical, 4x2¾x1½", 1961, brown plastic, six transistors, upper left front window dial with thumbwheel tuning, lower metal perforated grill area, rear swing-out stand, AM, bat **$50.00**

X16B, vertical, 1961, blue, seven transistors, upper right front round window dial with right side thumbwheel tuning, right front on/off/volume knob, large grill area with horizontal slots and lower left logo, swing handle, AM, bat.................... **$35.00**

X16G, vertical, 1961, green, seven transistors, upper right front round window dial with right side thumbwheel tuning, right front on/off/volume knob, large grill area with horizontal slots and lower left logo, swing handle, AM, bat.................... **$35.00**

X16N, vertical, 1961, brown/tan, seven transistors, upper right front round window dial with right side thumbwheel tuning, right front on/off/volume knob, large grill area with horizontal slots and lower left logo, swing handle, AM, bat **$35.00**

X17B, vertical, 5¾x4x1½", 1960, blue plastic, eight transistors, upper front window dial with right side thumbwheel tuning, right side thumbwheel on/off/volume knob, large round metal perforated grill area, swing handle, AM, bat **$40.00**

X17N, vertical, 5¾x4x1½", 1960, brown plastic, eight transistors, upper front window dial with right side thumbwheel tuning, right side thumbwheel on/off/volume knob, large round metal perforated grill area, swing handle, AM, bat .. **$40.00**

X17R, vertical, 5¾x4x1½", 1960, red plastic, eight transistors, upper front window dial with right side thumbwheel tuning, right side thumbwheel on/off/volume knob, large round metal perforated grill area, swing handle, AM, bat **$40.00**

X19A, vertical, 6¼x4x1¾", 1961, plastic, eight transistors, upper front horizontal, slide rule dial with thumbwheel tuning, lower perforated grill area with lower left logo, rear swing-out stand, leather handle, AM, bat **$30.00**

X19E, (bottom left) vertical, 6¼x4x 1¾", 1961, plastic, eight transistors, upper front horizontal slide rule dial with thumbwheel tuning, lower perforated grill area with lower left logo, rear swing-out stand, leather handle, AM, bat $30.00

X21W, vertical, 1961, plastic, six transistors, upper right front round dial knob, upper left front thumbwheel on/off/volume knob, lower metal perforated grill area with lower left logo with rhinestone decoration, AM, bat $45.00

X23B, vertical, 3⅝x2⅜x1⅛", 1962, blue plastic, six transistors, upper right front round dial knob, upper left front thumbwheel on/off/volume knob, lower patterned grill area, AM, bat........................ **$20.00**

X23E, vertical, 3⅝x2⅜x1⅛", 1962, black plastic, six transistors, upper right front round dial knob, upper left front thumbwheel on/off/volume knob, lower patterned grill area, AM, bat **$20.00**

X25E, vertical, 1962, black, six transistors, upper left front window dial with thumbwheel tuning, lower lattice grill area, AM, bat **$25.00**

X25J, vertical, 1962, jade, six transistors, upper left front window dial with thumbwheel tuning, lower lattice grill area, AM, bat **$25.00**

X26J, vertical, 1962, jade, seven transistors, upper left front round window dial with right side thumbwheel tuning, right side thumbwheel on/off/volume knob, large checkered grill area, AM, bat **$25.00**

X26W, vertical, 1962, white, seven transistors, upper left front round window dial with right side thumbwheel tuning, right side thumbwheel on/off/volume knob, large checkered grill area, AM, bat **$25.00**

X31A-1, horizontal, 5x8x3", 1962, gray leather, eight transistors, right front vertical slide rule dial, two knobs, left patterned grill area with upper left dial light and lower logo, leather handle, padded leather cover snaps in place to protect the radio front when not in use, AM, bat **$20.00**

X31B-1, horizontal, 5x8x3", 1962, blue leather, eight transistors, right front vertical slide rule dial, two knobs, left patterned grill area with upper left dial light and lower logo, leather handle, padded leather cover snaps in place to protect the radio front when not in use, AM, bat **$20.00**

X31E-1, horizontal, 5x8x3", 1962, black leather, eight transistors, right front vertical slide rule dial, two knobs, left patterned grill area with upper left dial light and lower logo, leather handle, padded leather cover snaps in place to protect the radio front when not in use, AM, bat **$20.00**

X31N "Ranger," horizontal, 5x7¾ x2½", 1962, brown leather, right front vertical slide rule dial, two knobs, left patterned grill area with upper left dial light and lower logo, leather handle, padded leather cover snaps in place to protect the radio front when not in use, AM, bat $20.00

X31N-1, horizontal, 5x7¾x2½", 1962, ginger leather, eight transistors, right front vertical slide rule dial, two knobs, left patterned grill area with upper left dial light and lower logo, leather handle, padded leather cover snaps in place to protect the radio front when not in use, AM, bat **$20.00**

X34B, vertical, 3⅝x2⅜x1", 1962, blue, six transistors, upper right front round dial knob, upper left thumbwheel on/off/volume knob, lower lattice grill area, AM, bat **$20.00**

X34E, vertical, 3⅝x2⅜x1", 1962, black, six transistors, upper right front round dial knob, upper left thumbwheel on/off/volume knob, lower lattice grill area, AM, bat **$20.00**

X35B, vertical, 3¾x2½x1⅛", 1962, blue plastic, six transistors, upper right front window dial with right side thumbwheel tuning, upper left front thumbwheel on/off/ volume knob, lower perforated grill area with lower left logo, AM, bat **$15.00**

X35E, vertical, 3¾x2½x1⅛", 1962, black plastic, six transistors, upper right front window dial with right side thumbwheel tuning, upper left front thumbwheel on/off/ volume knob, lower perforated grill area with lower left logo, AM, bat **$15.00**

X35N, vertical, 3¾x2½x1⅛", 1962, brown plastic, six transistors, upper right front window dial with right side thumbwheel tuning, upper left front thumbwheel on/off/ volume knob, lower perforated grill area with lower left logo, AM, bat $15.00

X36E, vertical, 4⅛x2¾x1½", 1962, black, six transistors, upper left front dial, right side thumbwheel knobs, lower lattice grill area, AM, bat **$20.00**

X36G, vertical, 4⅛x2¾x1½", 1962, green, six transistors, upper left front dial, right side thumbwheel knobs, lower lattice grill area, AM, bat **$20.00**

X37B, vertical, 4x3½x1½", 1962, blue leather, six transistors, right front window dial with right side thumbwheel tuning, right side thumbwheel on/off/volume knob, front perforated grill area, AM, bat **$25.00**

X37E, vertical, 4x3½x1½", 1962, black leather, six transistors, right front window dial with right side thumbwheel tuning, right side thumbwheel on/off/volume knob, front perforated grill area, AM, bat **$25.00**

X37S, vertical, 4x3½x1½", 1962, tan leather, six transistors, right front window dial with right side thumbwheel tuning, right side thumbwheel on/off/volume knob, front perforated grill area, AM, bat **$25.00**

X38EG, vertical, 5⅜x3¾x1⅝", 1962, black plastic, seven transistors, upper left front window dial with right side thumbwheel tuning, right side thumbwheel on/off/volume knob, large front perforated grill area, AM, bat $20.00

X38ES, vertical, 5⅜x3¾x1⅝", 1962, black plastic, seven transistors, upper left front window dial with right side thumbwheel tuning, right side thumbwheel on/off/volume knob, large front perforated grill area, AM, bat $20.00

X39E, horizontal, 4⅜x5¾x2", 1962, black leather, seven transistors, upper right front round window dial, lower metal grill area with vertical bars and lower left logo, leather handle, AM, bat $20.00

X39N, horizontal, 4⅜x5¾x2", 1962, brown leather, seven transistors, upper right front round window dial, lower metal grill area with vertical bars and lower left logo, leather handle, AM, bat.................... $20.00

X39S, horizontal, 4⅜x5¾x2", 1962, tan leather, seven transistors, upper right front round window dial, lower metal grill area with vertical bars and lower left logo, leather handle, AM, bat $20.00

X40E, vertical, 6⅝x4½x1¾", 1962, black plastic, eight transistors, foldback top, inner horizontal slide rule dial with thumbwheel tuning, lower perforated grill area, battery life indicator, right side strap, AM, bat $25.00

X40S, vertical, 6⅝x4½x1¾", 1962, sand plastic, eight transistors, foldback top, inner horizontal slide rule dial with thumbwheel tuning, lower perforated grill area, battery life indicator, right side strap, AM, bat $25.00

X41E, horizontal, 4¾x7½x2⅜", 1962, black leather, eight transistors, off-center horizontal slide rule dial, two right front thumbwheel knobs, left grill area with horizontal bars and battery life indicator, leather handle, AM, bat $15.00

X41E-1, horizontal, 4¾x7½x2⅜", 1962, black leather, eight transistors, off-center horizontal slide rule dial, two right front thumbwheel knobs, left grill area with horizontal bars and battery life indicator, leather handle, AM, bat $15.00

X41G, horizontal, 4¾x7½x2⅜", 1962, olive leather, eight transistors, off-center horizontal slide rule dial, two right front thumbwheel knobs, left grill area with horizontal bars and battery life indicator, leather handle, AM, bat $15.00

X41G-1, horizontal, 4¾x7½x2⅜", 1962, olive leather, eight transistors, off-center horizontal slide rule dial, two right front thumbwheel knobs, left grill area with horizontal bars and battery life indicator, leather handle, AM, bat **$15.00**

X41N, horizontal, 4¾x7½x2⅜", 1962, brown leather, eight transistors, off-center horizontal slide rule dial, two right front thumbwheel knobs, left grill area with horizontal bars and battery life indicator, leather handle, AM, bat **$15.00**

X41N-1, horizontal, 4¾x7½x2⅜", 1962, brown leather, eight transistors, off-center horizontal slide rule dial, two right front thumbwheel knobs, left grill area with horizontal bars and battery life indicator, leather handle, AM, bat **$15.00**

X42E-1, horizontal, 1963, leather, 10 transistors, right front vertical two-band slide rule dial, two knobs, left grill area with horizontal bars, telescoping antenna, handle, AM, FM, bat **$15.00**

X47B, vertical, 5⅜x3¾x1⅝", 1962, blue plastic, seven transistors, upper left front round window dial with right side thumbwheel tuning, right side thumbwheel on/off/volume knob, large checkered grill area, AM, bat **$20.00**

X47E, vertical, 5⅜x3¾x1⅝", 1962, black plastic, seven transistors, upper left front round window dial with right side thumbwheel tuning, right side thumbwheel on/off/volume knob, large checkered grill area, AM, bat **$20.00**

X48E, horizontal, 4⅜x5¾x2", 1962, black leather, seven transistors, upper right front round window dial, lower grill area with vertical bars and lower left logo, battery life indicator, leather handle, AM, bat **$15.00**

X48N, horizontal, 4⅜x5¾x2", 1962, brown leather, seven transistors, upper right front round window dial, lower grill area with vertical bars and lower left logo, battery life indicator, leather handle, AM, bat **$15.00**

X49B, horizontal, 6¾x7⅞x3¼", 1962, blue leather, six transistors, large right front round dial over perforated grill area, lower right on/off/volume knob, handle, AM, bat **$15.00**

X49E, horizontal, 6¾x7⅞x3¼", 1962, black leather, six transistors, large right front round dial over perforated grill area, lower right on/off/volume knob, handle, AM, bat **$15.00**

X49N, horizontal, 6¾x7⅞x3¼", 1962, brown leather, six transistors, large right front round dial over perforated grill area, lower right on/off/volume knob, handle, AM, bat **$15.00**

X50B, horizontal, 6¾x7⅞x3¼", 1962, blue leather, six transistors, large right front round dial over perforated grill area, lower right on/off/volume knob, handle, AM, bat/AC **$15.00**

X50E, horizontal, 6¾x7⅞x3¼", 1962, black leather, six transistors, large right front round dial over perforated grill area, lower right on/off/volume knob, handle, AM, bat/AC **$15.00**

X50N, horizontal, 6¾x7⅞x3¼", 1962, brown leather, six transistors, large right front round dial over perforated grill area, lower right on/off/volume knob, handle, AM, bat/AC **$15.00**

X51N, horizontal, 5⅜x8½x3¼", 1962, brown leather, nine transistors, large right horizontal slide rule dial, two lower knobs, left perforated grill area, padded leather cover snaps in place to protect the radio front when not in use, leather handle, AM, bat **$15.00**

X53EG, vertical, 5⅜x3¾x1⅝", 1962, black plastic, seven transistors, upper left front window dial with right side thumbwheel tuning, right side thumbwheel on/off/volume knob, large front perforated grill area, AM, bat **$20.00**

X53ES, vertical, 5⅜x3¾x1⅝", 1962, black plastic, seven transistors, upper left front window dial with right side thumbwheel tuning, right side thumbwheel on/off/volume knob, large front perforated grill area, AM, bat **$20.00**

X54B, vertical, 3½x2¼x1", 1962, blue plastic, six transistors, upper right front window dial with right side thumbwheel tuning, upper left thumbwheel on/off/volume knob, lower metal perforated grill area, AM, bat **$15.00**

X54E, vertical, 3½x2¼x1", 1962, black plastic, six transistors, upper right front window dial with right side thumbwheel tuning, upper left thumbwheel on/off/volume knob, lower metal perforated grill area, AM, bat $15.00

X57E, vertical, 1965, charcoal, seven transistors, large upper front round dial, lower horizontal grill bars with logo, AM, bat **$20.00**

X57N, vertical, 1965, beige, seven transistors, large upper front round dial, lower horizontal grill bars with logo, AM, bat **$20.00**

X60E, vertical, 4¾x3½x1½", 1965, large upper front horizontal slide rule dial, lower perforated grill area with logo, AM, bat.................. **$15.00**

X61E, horizontal, 1965, black leather, nine transistors, upper right front round window dial, two knobs, vertical grill bars with lower left logo, leather handle, AM, bat **$15.00**

X61N, horizontal, 1965, brown leather, nine transistors, upper right front round window dial, two knobs, vertical grill bars with lower left logo, leather handle, AM, bat **$15.00**

X80N, horizontal, 1964, 10 transistors, upper front horizontal two-band slide rule dial, large lower grill area with horizontal bars and lower left logo, telescoping antenna, leather handle, AM, FM, bat **$15.00**

XP23DW, vertical, plastic, 10 transistors, upper front horizontal slide rule dial with right side thumbwheel tuning, right side thumbwheel on/off/volume knob, large metal perforated grill area, AM, bat $15.00

XP64CE, horizontal, leather, large right front round dial, lower right on/off/volume knob, left metal per-

forated grill area, leather handle, AM, bat $15.00

XT18B, horizontal, 1961, six transistors, upper right front large round dial over lattice grill area, left on/off/volume knob and logo, feet, handle, AM, bat **$15.00**

XT18S, horizontal, 1961, six transistors, upper right front large round dial over lattice grill area, left on/off/volume knob and logo, feet, handle, AM, bat **$15.00**

Nanola

6TP-106, vertical, 1960, six transistors, upper front round dial, upper right thumbwheel on/off/volume knob, lower perforated grill area with vertical lines and lower left logo, AM, bat **$30.00**

National

T-21, vertical, 4¼x2½x1¼", plastic, seven transistors, two upper front window dials – one MW, one SW – with thumbwheel tuning, lower metal perforated grill area, swing handle, MW, SW, bat **$30.00**

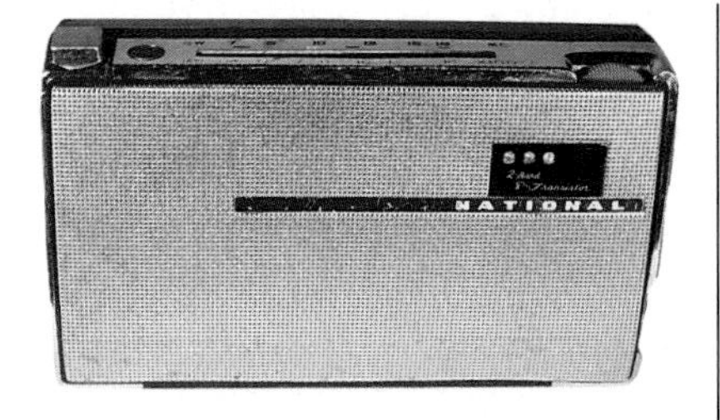

T-40, horizontal, 3⅝x6⅛x1⅜", plastic, eight transistors, top horizontal slide rule dial with top right thumbwheel tuning, lower right side thumbwheel on/off/volume knob, large front metal perforated grill area with three rhinestones, right side SW/MW switch, top left telescoping antenna, made in Japan, SW, MW, bat $40.00

T-98 "Portalarm," vertical/watch radio, 4x2½x1¼", plastic, seven transistors, upper right front window dial with thumbwheel tuning, upper left front Seiko alarm watch face, lower metal perforated grill area, rear stand, AM, bat $85.00

NEC

NT-61, horizontal, 1960, six transistors, right front thumbwheel dial and thumbwheel on/off/volume knob covered by a clear plastic panel, left grill area with textured horizontal bars, AM, bat **$50.00**

NT-620, horizontal, 2¾x4¼x1", 1960, plastic, six transistors, right front V-shaped dial opening with right side thumbwheel tuning, lower right front thumbwheel on/off/volume knob, left metal perforated grill area with circular indentations, AM, bat............ $50.00

NT-730, horizontal, 3⅛x5x1", plastic, seven transistors, upper right front horizontal dial with right side thumbwheel tuning, right side thumbwheel on/off/volume knob, lower front and rear metal perforated grill areas, made in Japan, AM, bat $25.00

Nobility

832N, vertical, 4¼x2½x1¼", plastic, eight transistors, upper right front square window dial with thumbwheel tuning, lower checkered grill area, AM, bat $10.00

Nordmende

"Condor," horizontal, 1963, nine transistors, upper right front round two-band dial overlaps checkered grill area, pushbuttons, telescoping antenna, handle, AM, FM, bat ... **$30.00**

"Mambo," horizontal, 1961, seven transistors, right front round two-band dial, left wedge-shaped lattice grill area, top pushbuttons and thumbwheel on/off/volume knob, handle, AM, LW, bat **$40.00**

"Mambino," horizontal, 1964, six transistors, upper right front round two-band dial overlaps checkered grill area, top pushbuttons and thumbwheel on/off/volume knob, handle, AM, LW, bat.............. **$30.00**

"Stradella," horizontal, 6x9¾x3", plastic leatherette, upper front horizontal three-band slide rule dial with top right thumbwheel tuning, lower metal perforated grill area, three top pushbuttons and thumbwheel on/off/volume knob, telescoping antenna, handle, made in West Germany, bat $30.00

"Transita," horizontal, 1961, nine transistors, rounded case, upper right front round three-band dial overlaps checkered grill area, four pushbuttons, telescoping antenna, handle, AM, FM, SW, bat **$40.00**

"Transita Deluxe K-C," horizontal, 1965, nine transistors, upper front horizontal three-band slide rule dial with top right thumbwheel tuning, lower metal perforated grill area, five top pushbuttons and thumbwheel on/off/volume knob, telescoping antenna, handle, AM, FM, SW, bat **$30.00**

"Transita Export-C," horizontal, 1964, nine transistors, upper front horizontal four-band slide rule dial with top right thumbwheel tuning, lower metal perforated grill area, five top pushbuttons and thumbwheel on/off/volume knob, telescoping antenna, handle, AM, FM, SW, LW, bat **$30.00**

Norelco

L0X95T/62R, horizontal, 1961, seven transistors, right front round dial, lower thumbwheel on/off/volume knob, left perforated grill area, made in Holland, AM, bat **$30.00**

L1W32T/02G, horizontal, 1965, eight transistors, upper front horizontal three-band slide rule dial, lower perforated grill area with lower right logo, telescoping antenna, AM, FM, LW, bat **$25.00**

L1X75T/64R, horizontal, 1960, seven transistors, right front round dial, upper thumbwheel on/off/volume knob, left perforated grill area, made in Holland, AM, bat **$30.00**

L1X75T/64RA, horizontal, 3½ x 6¼ x 1½, 1959, black plastic, seven transistors, right front round dial, upper thumbwheel on/off/volume knob, left metal perforated grill area, made in Holland, AM, bat **$30.00**

L2W54T/54R, horizontal, 1965, 10 transistors, upper front horizontal four-band slide rule dial with thumbwheel tuning, thumbwheel on/off/volume knob, four right band pushbuttons, perforated grill area, telescoping antenna, AM, FM, SW, LW, bat **$30.00**

L2X97T, horizontal/clock radio, 1960, seven transistors, right front round dial knob, upper thumbwheel on/off/volume knob, left front round clock face, AM, bat **$30.00**

L3X09T/54, horizontal, 1962, seven transistors, upper front horizontal three-band slide rule dial with thumbwheel tuning, thumbwheel on/off/volume knob, top pushbuttons, handle, AM, 2SW, bat ... **$25.00**

L3X19T/97, horizontal, 1964, seven transistors, upper front horizontal three-band slide rule dial with thumbwheel tuning, thumbwheel on/off/volume knob, five top recessed pushbuttons, lower horizontal grill bars, handle, AM, 2SW, bat **$25.00**

L3X76T/07, horizontal, 1961, seven transistors, upper front horizontal two-band slide rule dial with thumbwheel tuning, thumbwheel on/off/volume knob, four top recessed pushbuttons, handle, AM, SW, bat **$25.00**

L3X86T, horizontal, 1960, seven transistors, upper front horizontal two-band slide rule dial with thumbwheel tuning, thumbwheel on/off/volume knob, four top recessed pushbuttons, handle, AM, SW, bat **$25.00**

L3X88T, horizontal, 1960, seven transistors, upper front horizontal two-band slide rule dial with thumbwheel tuning, thumbwheel on/off/volume knob, four top recessed pushbuttons, handle, AM, SW, bat **$25.00**

L3X95T/00E, horizontal, 1961, yellow, seven transistors, right front round three-band dial over large checkered grill area, handle, AM, Marine, SW, bat **$30.00**

L3X95T/00W, horizontal, 1961, orange, seven transistors, right front round three-band dial over large checkered grill area, handle, AM, Marine, SW, bat **$30.00**

L3X95T/00X, horizontal, 1961, blue, seven transistors, right front round three-band dial over large checkered grill area, handle, AM, Marine, SW, bat **$30.00**

L4X05T, horizontal, 1961, seven transistors, upper front horizontal four-band slide rule dial with thumbwheel tuning, thumbwheel on/off/volume knob, seven top pushbuttons, large lower grill area, handle, AM, 3SW, bat **$35.00**

L4X25T, horizontal, 1964, seven transistors, upper front horizontal four-band slide rule dial with thumbwheel tuning, thumbwheel on/off/volume knob, seven top pushbuttons, large lower grill area, handle, AM, 3SW, bat **$35.00**

L4X95T, horizontal, 1960, brown leatherette with cream plastic front, seven transistors, upper front horizontal four-band slide rule dial with thumbwheel tuning, thumbwheel on/off/volume knob, seven top pushbuttons, large lower grill area, handle, AM, 3SW, bat **$35.00**

North American

876, horizontal, 2⅝x4½x1⅜", plastic, eight transistors, upper right front window dial with upper right side thumbwheel tuning, lower right side thumbwheel on/off/volume knob, oval grill area with horizontal bars, top left strap, made in Hong Kong, AM, bat $10.00

1400, horizontal, 1965, 14 transistors, upper front horizontal two-band slide rule dial, right tuning knob and AM/FM switch, left oval perforated grill area, telescoping antenna, handle, AM, FM, bat **$15.00**

Norwood

MN-1000, horizontal, 1965, 10 transistors, upper front horizontal two-band slide rule dial with thumbwheel tuning, large lower perforated grill area, telescoping antenna, handle, AM, FM, bat **$15.00**

NA-1200, horizontal, 1964, 12 transistors, right front dial with upper FM and lower AM windows, right side knob, left perforated grill area, telescoping antenna, handle, AM, FM, bat **$15.00**

NM-600 "HiFi Deluxe," vertical, 1964, six transistors, upper right front window dial with right side thumbwheel tuning, lower perforated grill area, AM, bat **$15.00**

NM-800, horizontal, 1964, eight transistors, upper front horizontal dial with right side thumbwheel tuning, left thumbwheel volume knob, large lower grill area with horizontal slots, AM, bat **$15.00**

NS-901, vertical, 1964, nine transistors, upper right front window dial with right side thumbwheel tuning, lower grill area with horizontal slots, AM, bat **$15.00**

NT-602, horizontal, 1964, six transistors, upper right front round window dial with upper right side thumbwheel tuning, lower right side thumbwheel on/off/volume knob, left perforated grill area, AM, bat **$20.00**

Nuvox

"Boy's Radio," vertical, 4⅛x2⅝x1¼", plastic, two transistors, upper right front window dial with right side thumbwheel tuning, left side thumbwheel on/off/volume knob, lower metal perforated grill area, made in Japan, AM, bat........ **$30.00**

Oldsmobile

"Trans-Portable," vertical car/portable radio, 6¾x3½x1¾", 1958, plastic, made to use in or out of the car, chrome front with upper round

dial knob, lower vertical grill bars, handle, made for Oldsmobile by Delco, AM $135.00

Olson

RA-315, vertical, 1960, four transistors, upper left front round dial with thumbwheel tuning, right side thumbwheel knob, lower lattice grill area with lower right logo, AM, bat **$40.00**

RA-347, horizontal, 1961, six transistors, right front window dial with right side thumbwheel tuning, right side thumbwheel on/off/volume knob, left perforated grill area, AM, bat **$25.00**

Olympic

447, horizontal, 1957, four transistors, top dial and on/off/volume knobs, large front checkered grill area, handle, AM, bat **$135.00**

666, horizontal, 1959, six transistors, off-center large round dial knob overlaps left grill area with diamond cut-outs, top right thumbwheel on/off/volume knob, AM, bat **$30.00**

766, vertical, 1959, leather, six transistors, upper left large round dial knob overlaps lower perforated grill area, top right thumbwheel on/off/volume knob, leather handle, AM, bat **$30.00**

768, horizontal, 1959, six transistors, right front round dial knob, lower right thumbwheel on/off/volume knob, left checkered grill area, handle, AM, bat **$35.00**

770, horizontal, 1959, six transistors, right front round dial knob, lower right on/off/volume window with lower right side thumbwheel knob, left grill area with lower left Olympic torch logo, AM, bat **$40.00**

771, horizontal, 1959, six transistors, right front round dial knob, upper right thumbwheel on/off/volume knob, "wishbone" grill decoration, handle, AM, bat **$60.00**

777, vertical, 1960, six transistors, upper front round window dial with thumbwheel tuning, lower perforated grill area with lower right Olympic torch logo, AM, bat **$35.00**

778, vertical, 1962, six transistors, upper right front round dial with right side thumbwheel tuning, top left thumbwheel on/off/volume knob, lower perforated grill area, AM, bat **$30.00**

779, vertical, 1962, six transistors, upper right front window dial with right side thumbwheel tuning, upper left front on/off/volume window with left side thumbwheel knob, lower round perforated grill area, lower right Olympic torch logo, AM, bat **$30.00**

781, vertical, 1962, six transistors, upper right front wedge-shaped window dial with right side thumbwheel tuning, top left thumbwheel on/off/volume knob, lower perforated grill area, AM, bat **$30.00**

808, horizontal, 1960, eight transistors, upper right front dial, lower right on/off/volume knob, large lattice grill area, handle, AM, bat **$25.00**

859, vertical, 1960, leather, eight transistors, upper front window dial with top thumbwheel tuning, top thumbwheel on/off/volume knob, lower perforated grill area, AM, bat **$35.00**

860, horizontal, 3½x5½x1½", 1963, plastic, eight transistors, upper right front round dial with right side thumbwheel tuning overlaps large metal perforated grill area, right side

thumbwheel on/off/volume knob, made in Japan, AM, bat $25.00

861, vertical, 1963, eight transistors, upper right front window dial with right side thumbwheel tuning, left side thumbwheel on/off/volume knob, perforated grill area with center Olympic torch logo, AM, bat **$30.00**

862, horizontal, 1964, eight transistors, upper right front window dial with upper right side thumbwheel tuning, lower right side thumbwheel on/off/volume knob, large perforated grill area, AM, bat **$20.00**

1063, horizontal, 1965, 10 transistors, upper right front window dial with right side thumbwheel tuning, lower right thumbwheel on/off/volume knob, left round perforated grill area, Olympic torch logo, AM, bat **$25.00**

1100 "High Fidelity," horizontal, 1963, 11 transistors, two upper front horizontal slide rule dials – one AM/SW, one FM – lower perforated grill area, telescoping antenna, handle, AM, FM, SW, bat **$20.00**

1200 "High Fidelity," horizontal, 1963, 12 transistors, two upper front horizontal slide rule dials – one AM, one FM – lower perforated grill area with two knobs, two telescoping antennas, handle, AM, FM, bat **$20.00**

CT999, horizontal/clock radio, 1963, six transistors, upper right front window dial over large perforated grill area with Olympic torch logo, right side thumbwheel tuning and on/off/volume knobs, left clock face, AM, bat **$20.00**

O.M.G.S.

700, vertical, 4¼x2½x1⅜", plastic, seven transistors, upper right front window dial with right side thumbwheel tuning, left side thumbwheel on/off/volume knob, large plastic checkered grill area, made in Hong Kong, AM, bat $15.00

TRN-8023 "Suburbia," horizontal, 3½x5¾x1⅜", plastic, 12 transistors, upper right front corner round dial knob, lower right side thumbwheel on/off/volume knob, left metal perforated grill area, made in Japan, AM, bat $35.00

Orion

JT-602 "Signal-Radio," vertical, 4¼x 2½x1¼", plastic, upper left front round dial knob, right side thumbwheel on/off/volume knob, lower metal perforated grill area, AM, bat $25.00

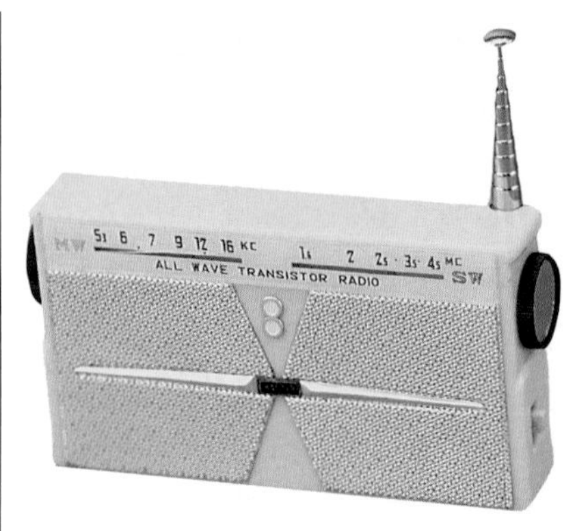

Master, horizontal, 3¾x7⅛x2", plastic, eight transistors, two upper front horizontal slide rule dials, right and left side knobs, right side switch, divided front metal perforated grill area with center logo, telescoping antenna, AM, SW, bat $35.00

Oritone

TR-107, vertical, 4¼x2½x1¼", plastic, 10 transistors, upper front window dial over large metal perforated grill area, right side thumbwheel tuning, top left thumbwheel

on/off/volume knob, made in Japan, AM, bat $30.00

OTL

10TP-904, horizontal, 10 transistors, right front dial with thumbwheel tuning, top left thumbwheel on/off/volume knob, large grill area with vertical slots and "starburst" decoration, bat $25.00

Oxford

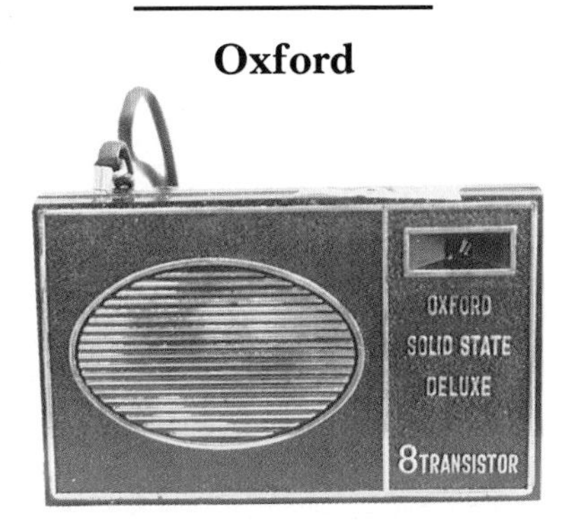

876, horizontal, 2⅝x4½x1⅜", plastic, eight transistors, upper right front window dial with right side thumbwheel tuning, right side thumbwheel on/off/volume knob, left oval grill area with horizontal bars, top left strap, AM, bat $10.00

1296 "XII," horizontal, 3¼x6⅛x1½", plastic, 12 transistors, upper right horizontal slide rule dial with right side thumbwheel tuning, right side thumbwheel on/off/volume knob, large oval metal perforated grill area, made in Japan, AM, bat $30.00

Packard Bell

6RT1, horizontal, 1958, leather, five transistors, right front round dial knob, right side thumbwheel on/off/volume knob, checkered grill cut-outs with lower left logo, handle, AM, bat **$60.00**

6RT-2, horizontal, 3¼x6½x1⅜", 1959, available in turquoise, red, chalk white, or ebony/white plastic, five transistors, right front round dial knob, left lattice grill area, AM, bat **$60.00**

6RT-6, vertical, 4x2½x1¼", 1964, six transistors, upper front dial with right side thumbwheel tuning, right side thumbwheel on/off/volume knob, lower metal perforated grill area with lower right logo, made in Japan, AM, bat $25.00

6RT-7, vertical, 1964, six transistors, upper front dial with thumbwheel tuning, lower perforated grill area with lower right logo, made in Japan, AM, bat **$25.00**

12RT1, horizontal, 1964, 12 transistors, upper left front horizontal two-band slide rule dial, three knobs, lower grill area, telescoping antenna, AM, FM, bat **$20.00**

Panasonic

R-8, horizontal/table, 1964, available in red, blue, white, or black, six transistors, right front horizontal slide rule dial, two knobs, left grill area with lower left logo, feet, AM, bat **$15.00**

R-102, vertical, 1964, six transistors, upper front round dial knob, right side thumbwheel on/off/volume knob, horizontal grill bars with lower logo, AM, bat **$15.00**

R-103, vertical, 1964, available in black, white, or blue, six transistors, upper front round dial knob, right side thumbwheel on/off/volume knob, lower grill area with center logo, AM, bat **$15.00**

R-109, horizontal, 1964, leather, nine transistors, upper right front horizontal slide rule dial, two knobs, left perforated grill area with lower left logo, handle, AM, bat **$15.00**

R-140 "Super Sensitive," horizontal, 1964, leather, six transistors, right front round dial overlaps lattice grill area, left side thumbwheel on/off/volume knob, leather handle, AM, bat **$15.00**

R-147, horizontal, 1965, leather, seven transistors, right front round dial overlaps lattice grill area, leather handle, AM, bat **$15.00**

R-505 "Super Sensitive," horizontal, 1964, seven transistors, right front window dial with right side thumbwheel tuning, top left thumbwheel on/off/volume knob, perforated grill area with lower left logo, AM, bat **$15.00**

R-607, vertical, 3½x2½x1", 1965, seven transistors, upper left front window dial with thumbwheel tuning, lower perforated grill area with lower right logo, AM, bat **$15.00**

R-1000 "Radar Matic," horizontal, 1965, 10 transistors, right front round signal-seeking dial with thumbwheel tuning, top left thumbwheel knob, perforated grill area with left logo, AM, bat **$30.00**

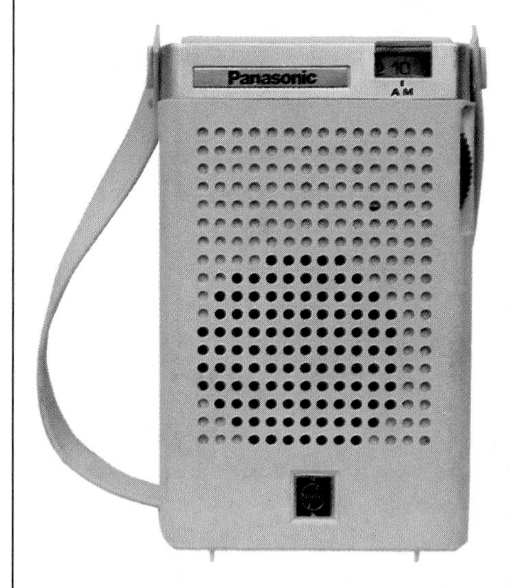

R-1029, vertical, 4½x3x1⅜", plastic, upper right window dial with right side thumbwheel tuning, left side thumbwheel on/off/volume knob,

circular grill cut-outs, vinyl strap, made in Japan, AM, bat $10.00

R-1070, vertical, 4½x2¾x1½", plastic, upper front horizontal slide rule dial with right side thumbwheel tuning, left side thumbwheel on/off/volume knob, circular grill cut-outs, vinyl strap, made in Taiwan, AM, bat **$10.00**

R-1082, vertical, 3¾x2½x1¼", plastic, upper right front window dial with thumbwheel tuning, lower checkered grill area with lower left logo, AM, bat **$15.00**

R-1157 "Super 7," horizontal, leather, right front round dial, left metal perforated grill area with lower left logo, handle, AM, bat $15.00

RF-811, vertical, 1965, eight transistors, upper front horizontal two-band slide rule dial with thumbwheel tuning, lower perforated grill area with lower left logo, telescoping antenna, AM, FM, bat **$15.00**

RF-815, horizontal, 1964, eight transistors, top horizontal two-band slide rule dial, front perforated grill area, telescoping antenna, AM, FM, bat **$20.00**

RF-820, horizontal, 1964, nine transistors, top horizontal two-band slide rule dial with thumbwheel tuning, right side switch, large front perforated grill area with lower left logo, telescoping antenna, AM, FM, bat **$20.00**

RF-835, horizontal, 7x9x3", 1965, leather, nine transistors, upper front horizontal two-band slide rule dial, lower perforated grill area, telescoping antenna, handle, AM, FM, bat **$15.00**

RL-112, horizontal/table, 1964, six transistors, right front round dial, lower on/off/volume knob, left grill area with horizontal bars, handle, feet, AM, bat **$10.00**

T-7, vertical, 1963, seven transistors, upper front horizontal slide rule dial with right side thumbwheel tuning, right side thumbwheel on/off/volume knob, lower grill area with horizontal slots and lower left logo, AM, bat $30.00

T-13P, vertical, 1964, six transistors, upper front horizontal slide rule dial with thumbwheel tuning, thumbwheel on/off/volume knob, lower grill area with lower right logo, AM, bat **$20.00**

T-33, horizontal, 1964, nine transistors, upper front horizontal two-band slide rule dial with thumbwheel tuning, upper left volume window with left side thumbwheel knob, large perforated grill area, band switch, telescoping antenna, handle, AM, FM, bat **$25.00**

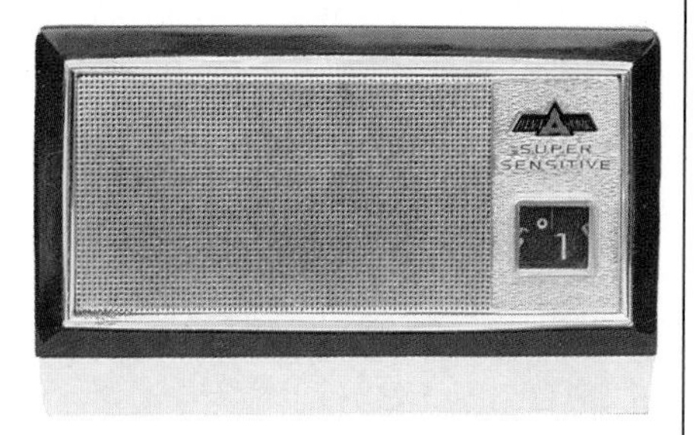

T-50 "Super Sensitive," horizontal, 3½x5¾x1½", 1962, plastic, six transistors, lower right front window dial with right side thumb-wheel tuning, top left thumbwheel on/off/volume knob, metal perforated grill area, made in Japan, AM, bat $25.00

T-50AA "Super Sensitive," horizontal, 3½x6x1¾", 1964, seven transistors, upper right front window dial with right side thumbwheel tuning, top left thumbwheel on/off/volume knob, metal perforated grill area, left side H/L switch, made in Japan, AM, bat **$25.00**

T-53, vertical, 3½x2½x1¼", 1963, plastic, six transistors, upper front horizontal slide rule dial with right side thumbwheel tuning, right side thumbwheel on/off/volume knob, lower perforated grill area with lower left logo, AM, bat $20.00

T-59, vertical, 1964, nine transistors, upper front horizontal slide rule dial with right side thumbwheel tuning, right side thumbwheel on/off/volume knob, lower perforated grill area with lower left logo, AM, bat **$15.00**

T-70M, horizontal, 1962, eight transistors, large right front horizontal two-band slide rule dial with thumbwheel tuning, thumbwheel on/off/volume knob, right side band switch, horizontal grill bars, telescoping antenna, handle, AM, SW, bat **$25.00**

T-70U, horizontal, 1962, eight transistors, large right front horizontal two-band slide rule dial with thumbwheel tuning, thumbwheel on/off/ volume knob, right side band switch, horizontal grill bars, telescoping antenna, handle, AM, SW, bat **$25.00**

T-81, horizontal, 1964, nine transistors, top horizontal two-band slide rule dial with thumbwheel tuning, thumbwheel on/off/volume knob, right side band switch, oval perforated grill area, telescoping antenna, AM, FM, bat **$25.00**

T-81H, horizontal, 1964, nine transistors, top horizontal three-band slide rule dial with thumbwheel tuning, thumbwheel on/off/volume knob, right side band switch, oval perforated grill area, telescoping antenna, AM, FM, SW, bat **$25.00**

T-92 "Portalarm," vertical watch/radio, 4x2½x1¼", 1962, plastic, six transistors, right front thumbwheel dial, left side thumbwheel volume knob, top right on/off/auto switch, left Seiko watch face, metal perforated grill area with lower left logo, made in Japan, AM, bat......... $80.00

T-100M, horizontal, 1965, 12 transistors, upper front horizontal four-band slide rule dial, large lower lattice grill area with lower right logo, telescoping antenna, handle, AM, FM, SW, LW, bat **$25.00**

T-601, vertical, 3½x2¼x1", 1964, plastic, upper left front window dial with thumbwheel tuning, large metal perforated grill area with lower right logo, rear fold-out stand, AM, bat........... $30.00

T-745, horizontal/table, 1964, wood, 12 transistors, lower slanted horizontal four-band slide rule dial, two knobs, large upper grill area, AM, FM, 2SW, bat **$25.00**

Peerless

10T-2SP "Twin Speaker," horizontal, 1964, 10 transistors, upper center front horizontal slide rule dial with thumbwheel tuning, right and left grill areas with horizontal slots, twin speakers, AM, bat **$20.00**

630, vertical, 1965, six transistors, upper right front window dial with right side thumbwheel tuning, top left thumbwheel on/off/volume knob, lower grill area with horizontal slots, AM, bat **$15.00**

707, horizontal, 1964, seven transistors, upper front horizontal dial, right and left knobs, lower oval grill area with center "707" emblem, handle, AM, bat.................... **$20.00**

777, vertical, 1965, seven transistors, upper right front round window dial with right side thumbwheel tuning, top left thumbwheel on/off/volume knob, lower round grill area, AM, bat **$15.00**

820, vertical, 1965, eight transistors, upper right front window dial with right side thumbwheel tuning, left side thumbwheel on/off/volume knob, lower perforated grill area, AM, bat **$20.00**

830, vertical, 1965, eight transistors, upper left front round dial, lower round grill area, AM, bat....... **$20.00**

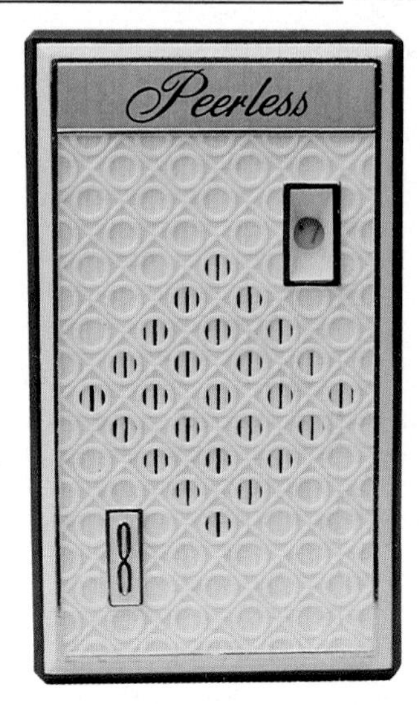

855, vertical, 4¼x2½x1¼", plastic, eight transistors, upper right front round window dial over front panel with diamond and circular cutouts, right side thumbwheel tuning, left side thumbwheel on/off/volume knob, made in Hong Kong, AM, bat $20.00

880, horizontal, 1965, eight transistors, right front round window dial with thumbwheel tuning, lower right thumbwheel on/off/volume knob, left grill area with horizontal slots, swing handle, AM, bat **$15.00**

990, horizontal, 1965, leather, nine transistors, upper front horizontal two-band slide rule dial, right and left knobs, lower oval perforated grill area with center "990" emblem, telescoping antenna, handle, AM, FM, bat **$20.00**

1030 "Hi-Fi," vertical, 1965, ten transistors, upper front horizontal slide rule dial with thumbwheel tuning, lower grill area with center logo, strap, AM, bat **$15.00**

1033, horizontal, 1965, 10 transistors, upper front horizontal three-band slide rule dial with lower right tuning knob, left grill area with horizontal slots, telescoping antenna, handle, AM, Marine, SW, bat **$20.00**

1200, vertical, 1965, 12 transistors, upper right front thumbwheel dial, top left thumbwheel on/off/volume knob, large grill area with center logo, strap, AM, bat **$15.00**

1333, horizontal, 1965, leather, 13 transistors, upper slanted horizontal three-band slide rule dial, two lower right knobs, left grill area with horizontal slots, telescoping antenna, leather handle, AM, FM, SW, bat **$25.00**

FM-90, vertical, 1965, nine transistors, upper front round two-band dial with thumbwheel tuning overlaps large lower grill area, telescoping antenna, AM, FM, bat...... **$15.00**

Penncrest

1130, horizontal, 2⅝x4½x1⅛", plastic, six transistors, upper right front window dial with upper right side thumbwheel tuning, lower right side thumbwheel on/off/volume knob, large metal grill area with horizontal slots and lower left logo, made in Japan, AM, bat $25.00

1132, horizontal, 2½x4¼x1", plastic, six transistors, upper right front window dial with upper right side thumbwheel tuning, lower right side thumbwheel on/off/volume knob, large metal perforated grill area with lower left logo, AM, bat **$20.00**

1631, horizontal, 1964, six transistors, upper front horizontal slide rule dial, right and left knobs, large lower perforated grill area with right tone knob and left logo, handle, AM, bat................. **$15.00**

1871, horizontal, 1964, 10 transistors, two right front dials – upper AM, lower FM – left perforated grill area with upper left logo, telescoping antenna, strap, AM, FM, bat **$25.00**

1896, horizontal, 1963, leather, 12 transistors, two upper front round dials – right AM, left FM – thumbwheel volume and tone knobs, band switch, perforated grill area with logo, telescoping antenna, leather handle, AM, FM, bat **$25.00**

1991, horizontal, 1964, 12 transistors, upper front horizontal dial, right and left knobs, lower perforated grill area with lower left logo, two telescoping antennas, handle, AM, FM, SW, bat **$25.00**

Penney's

41-8TM-360, horizontal, 3¾x6½x1½", plastic, eight transistors, upper front wrap-over dial, right side tuning knob, left side volume knob, metal perforated grill area with "V" decoration, AM, bat **$35.00**

1140, square, plastic, right side thumbwheel tuning and on/off/ volume knobs, front round metal perforated grill area, top vinyl strap, AM, bat $15.00

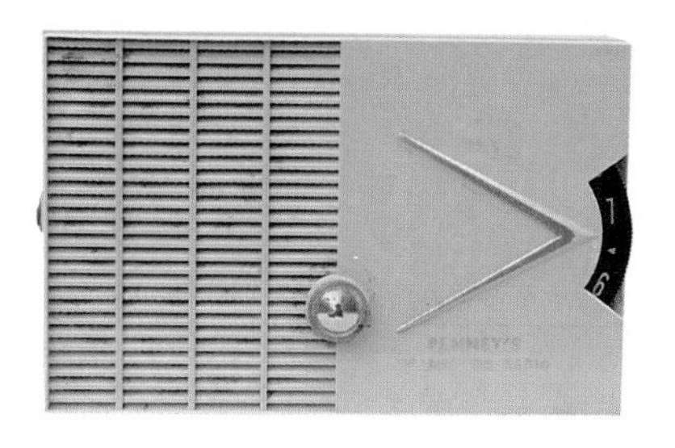

RP-1-124, horizontal, 3⅝x5⅞x1⅝", 1958, plastic, right front thumbwheel dial with V-shaped pointer, center on/off/volume knob, left lattice grill area, made in USA, AM, bat $40.00

Perdio

PC16, horizontal, 4⅛x6⅝x1¾", 1961, leather, seven transistors, left front round brass dial knob, top on/off/ volume knob under leather handle, cloth grill area, lower left lion logo, made in England, AM, bat **$30.00**

Pet

"Boy's Radio," vertical, 3x2¼x1¼", plastic, two transistors, upper right front window dial with right side thumbwheel tuning, left side thumbwheel on/off/volume knob, round metal perforated grill area, AM, bat $35.00

Petite

6G620, vertical, plastic, six transistors, upper left front window dial with left side thumbwheel tuning, right side thumbwheel on/off/volume knob, lower metal perforated grill area with lower right triangular logo, AM, bat **$25.00**

NTR-120, vertical, 1961, six transistors, upper left front window dial with top left thumbwheel tuning, top right thumbwheel on/off/volume knob, lower perforated grill area with lower left logo, AM, bat **$35.00**

NTR-150, vertical, 1961, six transistors, upper front window dial with thumbwheel tuning, lower perforated grill area with oriental design, AM, bat **$40.00**

NTR-800, horizontal, 1964, eight transistors, upper front horizontal two-band slide rule dial with right side thumbwheel tuning, perforated grill area with lower right band switch, AM, SW, bat **$25.00**

Philco

NT-600BKG, vertical, $3^3/_4$x$2^1/_4$x1", 1964, plastic, six transistors, upper right front window dial with right side thumbwheel tuning, top left thumbwheel on/off/volume knob, metal perforated grill area with lower left logo, AM, bat $20.00

NT-601BK, horizontal, 1965, six transistors, right front round dial knob, lower on/off/volume knob, left grill area, handle, AM, bat **$15.00**

NT-802WHG, vertical, 1964, eight transistors, upper front round dial over large perforated grill area, right side thumbwheel tuning knob, right side thumbwheel on/off/volume knob, AM, bat **$20.00**

NT-807, horizontal, 1965, eight transistors, upper front horizontal dial with thumbwheel tuning, lower two-section perforated grill area, swing handle, AM, bat **$20.00**

NT-808, horizontal, 1965, leather, eight transistors, upper front horizontal slide rule dial, two knobs, lower grill area with horizontal bars, strap, AM, bat **$15.00**

NT-814BKG, vertical, 1965, eight transistors, two upper front round dials – one AM, one FM – lower perforated grill area with lower left logo, telescoping antenna, made in Japan, AM, FM, bat **$20.00**

NT-815BK, horizontal, 1965, leather, ten transistors, off-center vertical three-band slide rule dial, three right knobs, left perforated grill area, telescoping antenna, leather handle, AM, 2SW, bat **$20.00**

NT-900, vertical, 1964, nine transistors, upper front round dial over large perforated grill area, right side knob, strap, AM, bat **$20.00**

NT-903BK, horizontal, 1965, nine transistors, right front round two-band dial overlaps horizontal grill bars, telescoping antenna, handle, AM, FM, bat **$20.00**

NT-906BKG, horizontal, 1964, nine transistors, step-back top, two off-center round dials – one FM, one AM – right tone and off/on/volume knobs, right side thumbwheel tuning knob, left perforated grill area, handle, AM, FM, bat **$20.00**

NT-912BK, horizontal, 1965, leather, ten transistors, large off-center round two-band dial, three right knobs, left grill area, telescoping antenna, leather handle, AM, FM, AC/bat................................. **$15.00**

NT-913BK, horizontal, 8x11x4½", 1965, leather, 11 transistors, two upper left front dials – one FM, one AM – three right knobs, horizontal grill bars, telescoping antenna, leather handle, AM, FM, bat **$15.00**

NT-1004, horizontal, 1965, leather, 10 transistors, upper front horizontal slide rule dial, three right knobs, left checkered grill area with lower left logo, telescoping antenna, leather handle, AM, FM, bat **$15.00**

T-4-124, horizontal, 1959, four transistors, right front round dial knob, center on/off/volume knob, left horizontal grill bars, AM, bat **$45.00**

T5-124 "500," horizontal, 3½x5⅞x 1¾", plastic, right front window dial over large metal perforated

grill area, right side thumbwheel tuning and on/off/volume knobs, AM, bat $30.00

T-6, horizontal, 6⅛x9¼x3", 1958, leather, six transistors, right front round dial knob, left side on/off/ volume knob, leather handle, AM, bat **$20.00**

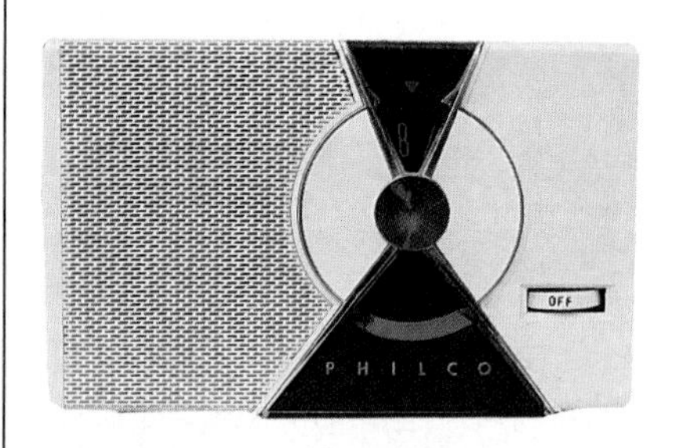

T-7-126, horizontal, 4¼x7x1¾", 1956, plastic, seven transistors, Philco's first transistor radio, two-tone front panel with off-center thumbwheel dial accessible through two triangular openings, lower right thumbwheel on/off/volume knob, left painted metal perforated grill area, AM, bat $65.00

T-9 "Trans World," horizontal, 11x 16½x6½", 1958, leather, nine transistors, fold-up front cover, inner right nine-band slide rule dial, left checkered grill area, four knobs, telescoping antenna, bat **$90.00**

T-50, vertical, $4\frac{5}{8}$x$3\frac{1}{8}$x$1\frac{7}{8}$", 1959, plastic, five transistors, upper front round dial knob, right side thumbwheel on/off/volume knob, lower horizontal grill bars, AM, bat **$30.00**

T51-124, vertical, $4\frac{5}{8}$x$3\frac{1}{8}$x$1\frac{7}{8}$", 1961, plastic, five transistors, upper front round dial knob, right side thumbwheel on/off/volume knob, lower horizontal grill bars, AM, bat $30.00

T52-124, horizontal, 1961, available in charcoal, ivory, or terra cotta, five transistors, right front recessed horizontal slide rule dial with right side thumbwheel tuning, center on/off/volume knob, horizontal grill bars, AM, bat $20.00

T-60, vertical, $5\frac{5}{8}$x$3\frac{1}{4}$x$1\frac{3}{4}$", 1959, available in gold/ebony or charcoal/ivory plastic, six transistors, upper front thumbwheel dial, right thumbwheel on/off/volume knob, lower horizontal grill bars, swing handle, AM, bat **$35.00**

T61-124, vertical, $4\frac{3}{8}$x$2\frac{5}{8}$x$1\frac{1}{4}$", plastic, upper right front window dial with right side thumbwheel tuning, left on/off/volume window with left side thumbwheel knob, lower checkered grill area, AM, bat $15.00

T-62, vertical, plastic, upper front thumbwheel dial in V-shaped indent, right thumbwheel on/off/volume knob, lower horizontal grill bars, swing handle, AM, bat **$35.00**

T-64, horizontal, 3¾x6x1½", 1963, available in brown or blue plastic, six transistors, upper right front round dial, lower on/off/volume knob, left horizontal grill bars with center logo, AM, bat **$15.00**

T-65, horizontal, 6½x8⅞x3", 1959, available in ivory/aqua or ivory/gold plastic, six transistors, right front dial knob, left horizontal grill bars, feet, rotatable antenna in handle, AM, bat **$25.00**

T-66AQ, vertical, 1961, six transistors, upper right front round dial with right side thumbwheel tuning, lower checkered grill area, AM, bat **$25.00**

T-67GP, horizontal, 1963, six transistors, right front window dial over large perforated grill area, right side thumbwheel knobs, AM, bat **$25.00**

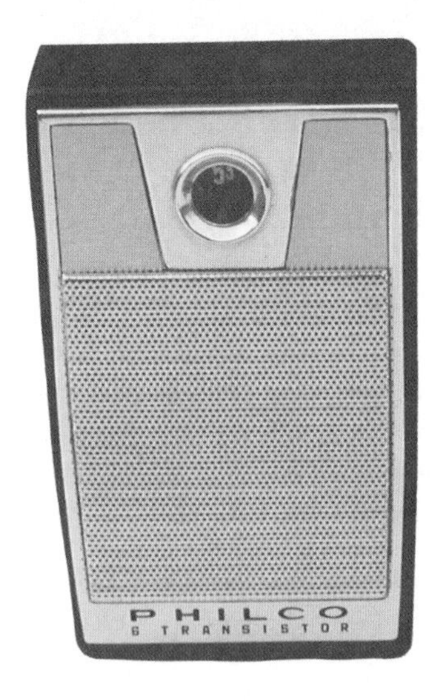

T-68BKG, vertical, 4x2½x1½", 1963, plastic, six transistors, upper front round window dial with thumbwheel tuning, lower metal perforated grill area, AM, bat $25.00

T-69WH, horizontal, 1964, six transistors, right front window dial with upper right side thumbwheel tuning, lower right side thumbwheel on/off/volume knob, large perforated grill area, AM, bat **$20.00**

T-70-124, vertical, 5½x3¼x1¾", 1961, plastic, seven transistors, upper front thumbwheel dial in V-shaped indent, right thumbwheel on/off/volume knob, lower horizontal grill bars, swing handle, AM, bat $35.00

T-74-124, horizontal, 1961, leather, seven transistors, off-center horizontal slide rule dial with right side thumbwheel tuning, leather handle, AM, bat **$25.00**

T-75-124, horizontal, $5\frac{1}{2}$x$7\frac{1}{8}$x$2\frac{1}{4}$", 1959, leather, seven transistors, off-center round thumbwheel dial overlaps metal perforated grill area, lower right thumbwheel on/off/volume knob, leather strap, AM, bat $30.00

T-76-124, horizontal, $5\frac{1}{2}$x$7\frac{1}{4}$x$2\frac{1}{2}$", 1960, leather, seven transistors, off-center round thumbwheel dial overlaps metal perforated grill area, lower right thumbwheel on/off/volume knob, leather strap, AM, bat $30.00

T-77-124, vertical, 1962, plastic, seven transistors, upper right front dial overlaps large grill area with horizontal slots, right side thumbwheel tuning, left front on/off/volume window with left side thumbwheel knob, AM, bat **$25.00**

T-77BK, vertical, 1962, seven transistors, upper right front dial overlaps large grill area with horizontal slots, right side thumbwheel tuning, left front on/off/volume window with left side thumbwheel knob, AM, bat **$25.00**

T-78-124, horizontal, $7\frac{1}{8}$x$9\frac{5}{8}$x4", 1959, leather, seven transistors, right front oblong window dial and knob, left side on/off/volume knob, round metal grill area with triangular perforations, handle, AM, bat $20.00

T-84BR, horizontal, 1963, leather, eight transistors, right front dial with lower tuning knob and upper on/off/volume knob, large perforated grill area with lower left logo, handle, AM, bat **$15.00**

T-90, vertical, 1963, nine transistors, upper front round dial knob over large grill area, right side on/off/volume knob, strap, AM, bat **$25.00**

T-500-124 "500," horizontal, $3\frac{1}{2}$x$5\frac{7}{8}$x$1\frac{3}{4}$", 1958, plastic, right front window dial over large metal perforated grill area, right side thumbwheel tuning and on/off/volume knobs, AM, bat **$30.00**

T-700-124 "700," horizontal, 6¾x9½x 4", 1958, leather, seven transistors, lower right front round dial knob overlaps metal grill area, left side on/off/volume knob, leather handle, AM, bat $25.00

T-701-124, horizontal, lower right front dial overlaps metal perforated grill area, left side on/off/volume knob, top "Scantenna" antenna/ handle, AM, bat $30.00

T-702-124 "VII," horizontal, 1960, leather, seven transistors, upper right front window dial with large lower tuning knob, sliding on/off switch next to tuning window, left side volume knob, vertical grill bars, leather handle, AM, bat **$25.00**

T-703-124 "VII," horizontal, 5⅜x9x 3¾", 1963, leather, seven transistors, upper right front window dial with large lower tuning knob, sliding on/ off switch next to tuning window, left side volume knob, vertical grill bars, leather handle, AM, bat **$25.00**

T-802-124, horizontal, 1961, leather, eight transistors, right round dial knob over large checkered grill area, leather strap, AM, bat **$20.00**

T-803-124, horizontal, 6⅞x9⅝x3¾", 1961, leather, eight transistors, right front round dial knob overlaps metal horizontal grill bars, left side on/off/volume knob, leather strap, AM, bat $20.00

T-804, horizontal, 1963, leather, eight transistors, right front round dial knob overlaps horizontal grill bars, leather strap, AM, bat ... **$20.00**

T-905-124, horizontal, 1962, nine transistors, right front two-band dial over large checkered grill area, AM/ FM pushbuttons, telescoping antenna, handle, AM, FM, bat ... **$25.00**

T-907, horizontal, 1963, available in brown or black, nine transistors, two upper front horizontal dial scales – one AM, one FM – large lower checkered grill area with AM/FM switch, telescoping antenna, handle, AM, FM, bat.................................. **$20.00**

T-908GY, horizontal, 1964, nine transistors, upper front horizontal two-band slide rule dial, top pushbuttons, lower checkered grill area, telescoping antenna, handle, AM, FM, bat........ **$25.00**

T-911, horizontal, 1963, nine transistors, upper front horizontal four-band slide rule dial, lower checkered grill area, pushbuttons, two telescoping antennas, handle, AM, FM, SW, LW, bat.................. **$25.00**

T1000-124, horizontal/clock radio, 8½x15½x5", 1960, plastic/metal, six transistors, modernistic design consists of three modules (right/left speakers and center clock face) swivel-mounted on base, thumbwheel tuning and on/off/volume knobs, AM, bat.................. $100.00

TC-47, horizontal/clock radio, 1960, leather, four transistors, right front round dial, center round grill area and on/off/volume knob, left round clock face, AM, bat................ **$30.00**

Plata

8R-34 "High Fidelity Deluxe," horizontal, 1965, leather, eight transistors, upper right front round dial, upper left on/off/volume knob over large perforated grill area, handle, AM, bat **$15.00**

8R-75 "High Fidelity," horizontal, 1965, leather, eight transistors, off-center vertical slide rule dial, three right knobs, left grill area with horizontal bars, leather handle, AM, bat **$15.00**

8S-31, horizontal, 1965, eight transistors, upper front horizontal two-band slide rule dial with right side thumbwheel tuning, lower lattice grill area, telescoping antenna, AM, SW, bat **$20.00**

9TA-370, horizontal, 1962, 10 transistors, upper front horizontal four-band slide rule dial with lower right tuning knob, thumbwheel tone and volume knobs, perforated grill area, telescoping antenna, handle, AM, 2SW, LW, bat..........................**$25.00**

10TF-530, horizontal, 4¾x7⅞x2⅛", plastic, upper front horizontal two-band dial with right tuning knob, top left thumbwheel on/off/volume knob, large front metal perforated grill area with center nameplate, right side AM/FM switch, telescoping antenna, made in Japan, AM, FM, bat $25.00

Polyrad

P-86, vertical, 1961, six transistors, upper left front round thumbwheel dial knob, upper right thumbwheel on/off/volume knob, lower perforated grill area, AM, bat **$25.00**

Princeton

WTC-610, vertical, 4½x2⅝x1¼", plastic, six transistors, upper right front window dial with right side thumbwheel tuning, left side thumbwheel on/off/volume knob, lower lattice grill area, top right vinyl strap, made in Okinawa, AM, bat.... $10.00

Raleigh

805, vertical, 1965, eight transistors, upper right front oval window dial with thumbwheel tuning, top left thumbwheel on/off/volume knob, lower oval grill area, AM, bat **$20.00**

828, vertical, 1965, eight transistors, upper right front round window dial with thumbwheel tuning, lower perforated grill area, AM, bat **$15.00**

885 "High Sensitivity," horizontal, 1965, eight transistors, right front dial, left grill area with horizontal bars, handle, AM, bat **$20.00**

1005, vertical, 1965, 10 transistors, upper right front oval window dial with right side thumbwheel tuning, top left thumbwheel on/off/volume knob, lower oval grill area, AM, bat **$20.00**

FM-925, vertical, 1965, nine transistors, two upper front window dials – one FM, one AM – with thumbwheel tuning, lower horizontal grill bars, telescoping antenna, strap, AM, FM, bat **$15.00**

Raytheon

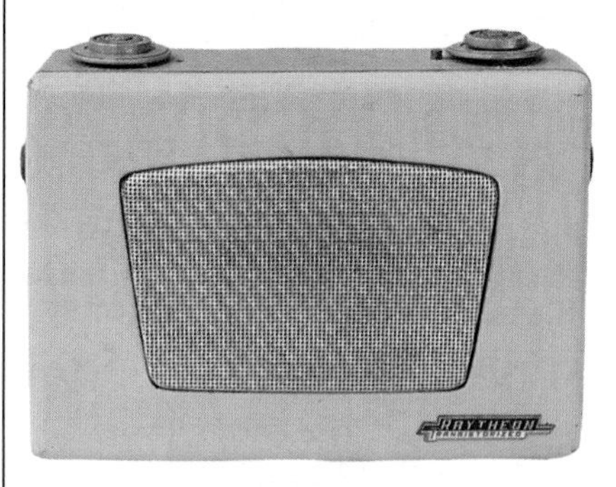

8TP-1, horizontal, 7x9¼x2¾", 1955, tan leather, eight transistors, Raytheon's first transistor radio, top controls – right dial knob, left on/off/volume knob – front and rear metal perforated grill areas, handle, AM, bat $125.00

8TP-2, horizontal, 7x9¼x2¾", 1955, brown leather, eight transistors, Raytheon's first transistor radio, top controls – right dial knob, left on/off/volume knob – front and rear metal perforated grill areas, handle, AM, bat **$125.00**

8TP-3, horizontal, 7x9¼x2¾", 1955, beige leather, eight transistors, Raytheon's first transistor radio, top controls – right dial knob, left on/off/volume knob – front and rear metal perforated grill areas, handle, AM, bat **$125.00**

8TP-4, horizontal, 7x9¼x2¾", 1955, red leather, eight transistors, Raytheon's first transistor radio, top controls – right dial knob, left on/off/volume knob – front and rear metal perforated grill areas, handle, AM, bat **$125.00**

T-100-1, horizontal, 3⅜x6⅜x2", 1956, black/yellow plastic, right front round brass dial over checkered grill area, thumbwheel on/off/volume knob, right side strap, AM, bat **$200.00**

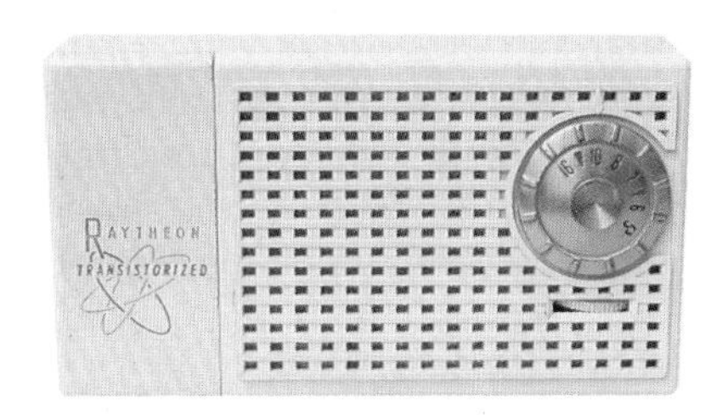

T-100-2, horizontal, 3⅜x6⅜x2", 1956, ivory/yellow plastic, right front round brass dial over checkered grill area, thumbwheel on/off/volume knob, right side strap, AM, bat $200.00

T-100-3, horizontal, 3⅜x6⅜x2", 1956, black/red plastic, right front round brass dial over checkered grill area, thumbwheel on/off/volume knob, right side strap, AM, bat $200.00

T-100-4, horizontal, 3⅜x6⅜x2", 1956, ivory/red plastic, right front round brass dial over checkered grill area, thumbwheel on/off/volume knob, right side strap, AM, bat **$200.00**

T-100-5, horizontal, 3⅜x6⅜x2", 1956, ivory/gray plastic, right front round brass dial over checkered grill area, thumbwheel on/off/volume knob, right side strap, AM, bat....... **$200.00**

T-150-1, horizontal, 3⅜x6⅜x2", 1956, black/yellow plastic, right front round brass dial over checkered grill area, thumbwheel on/off/volume knob, right side strap, AM, bat **$200.00**

T-150-2, horizontal, 3⅜x6⅜x2", 1956, ivory/yellow plastic, right front round brass dial over checkered grill area, thumbwheel on/off/volume knob, right side strap, AM, bat **$200.00**

T-150-3, horizontal, 3³⁄₈x6³⁄₈x2", 1956, black/red plastic, right front round brass dial over checkered grill area, thumbwheel on/off/volume knob, right side strap, AM, bat **$200.00**

T-150-4, horizontal, 3³⁄₈x6³⁄₈x2", 1956, ivory/red plastic, right front round brass dial over checkered grill area, thumbwheel on/off/volume knob, right side strap, AM, bat $200.00

T-150-5, horizontal, 3³⁄₈x6³⁄₈x2", 1956, ivory/gray plastic, right front round brass dial over checkered grill area, thumbwheel on/off/volume knob, right side strap, AM, bat....... **$200.00**

T-2500, horizontal, 9x12¹⁄₂x5³⁄₄", 1956, cloth covered with metal trim, seven transistors, top controls – right dial knob, left on/off/volume knob – front and rear metal perforated grill areas, two speakers, handle, AM, bat $100.00

RCA

1-BT-21 "Transicharg," horizontal, 1959, six transistors, can be used with matching battery charger unit, right front round dial knob, lower thumbwheel on/off/volume knob, left horizontal grill bars, AM, bat.
radio without charger **$35.00**
radio with charger **$125.00**

1-BT-24 "Transicharg," horizontal, 1959, six transistors, can be used with matching battery charger unit, right front round dial knob, lower thumbwheel on/off/volume knob, left horizontal grill bars, AM, bat.
radio without charger **$35.00**
radio with charger **$125.00**

1-BT-29 "Transicharg Super," horizontal, 1959, six transistors, can be used with matching battery charger unit, right front round dial knob, lower thumbwheel on/off/volume knob, left horizontal grill bars, AM, bat.
radio without charger **$35.00**
radio with charger **$125.00**

1-BT-32 "Transicharg Deluxe," horizontal, 3⁵⁄₈x7¹⁄₂x2", 1959, white/pink, seven transistors, can be used with matching battery charger unit, right front round dial knob, lower thumbwheel on/off/volume knob, left horizontal grill bars, swing handle, AM, bat.
radio without charger **$35.00**
radio with charger **$125.00**

1-BT-34 "Transicharg Deluxe," horizontal, 3⁵⁄₈x7¹⁄₂x2", 1959, white/

green, seven transistors, can be used with matching battery charger unit, right front round dial knob, lower thumbwheel on/off/volume knob, left horizontal grill bars, swing handle, AM, bat.
radio without charger **$35.00**
radio with charger **$125.00**

1-BT-36 "Transicharg Deluxe," horizontal, 3⅝x7½x2", 1959, gray/white, seven transistors, can be used with matching battery charger unit, right front round dial knob, lower thumbwheel on/off/volume knob, left horizontal grill bars, swing handle, AM, bat.
radio without charger $35.00
radio with charger $125.00

1-BT-41, horizontal, 1958, antique white leather, six transistors, upper right front round dial knob, top right thumbwheel on/off/volume knob, perforated grill area, leather handle, AM, bat **$25.00**

1-BT-46, horizontal, 1958, charcoal leather, six transistors, upper right front round dial knob, top right thumbwheel on/off/volume knob, perforated grill area, leather handle, AM, bat **$25.00**

1-BT-48, horizontal, 1958, russet leather, six transistors, upper right front round dial knob, top right thumbwheel on/off/volume knob, perforated grill area, leather handle, AM, bat **$25.00**

1-BT-58 "Globe Trotter," horizontal, 7x11x4", 1959, leather, seven transistors, upper front horizontal slide rule dial, right and left side knobs, lower perforated grill area, leather handle, AM, bat **$25.00**

1-MBT-6 "Strato-World," horizontal, 1959, leather, nine transistors, fold-up front with world map, inner horizontal seven-band slide rule dial, telescoping antenna, handle, bat **$75.00**

1-RG-11, vertical, 7x4⅛x2⅛", 1962, black/gray plastic, six transistors, upper front round dial knob, lower right side on/off/volume knob, vertical grill bars with lower right Nipper and RCA logos, swing handle, AM, bat $20.00

1-RG-14, vertical, 7x4⅛x2⅛", 1962, brown/white plastic, six transistors, upper front round dial knob, lower right side on/off/volume knob, vertical grill bars with lower right Nipper and RCA logos, swing handle, AM, bat $20.00

1-RG-15, vertical, 7x4⅛x2⅛", 1962, two-tone green plastic, six transistors, upper front round dial knob, lower right side on/off/volume knob, vertical grill bars with lower right Nipper and RCA logos, swing handle, AM, bat **$20.00**

1-RG-41, horizontal, 1963, six transistors, upper right front round window dial with thumbwheel tuning, left perforated grill area, handle, AM, bat **$20.00**

1-RH-10, vertical, 3⅞x2½x1⅛", 1961, plastic, six transistors, upper front round dial knob, right side thumbwheel on/off/volume knob, lower lattice grill area, made in USA, AM, bat $20.00

1-RH-11, vertical, 3⅞x2½x1⅛", 1961, plastic, six transistors, upper front round dial knob, right side thumbwheel on/off/volume knob, lower lattice grill area, made in USA, AM, bat **$20.00**

1-RJ-19, vertical, 4x2½x1¼", 1961, plastic, six transistors, upper front round dial knob, right side thumbwheel on/off/volume knob, lattice grill area, AM, bat **$20.00**

1-T-1DJ, vertical, 7x4x2", 1960, plastic, six transistors, upper front round dial, lower right side on/off/volume knob, lower vertical grill bars with lower right Nipper and RCA logos, swing handle, AM, bat **$25.00**

1-T-1E, vertical, 7x4x2", 1960, plastic, six transistors, upper front round dial, lower right side on/off/volume knob, lower vertical grill bars with lower right Nipper and RCA logos, swing handle, AM, bat $25.00

1-T-1LE, vertical, 7x4x2", 1960, plastic, six transistors, upper front round dial, lower right side on/off/volume knob, lower vertical grill bars with lower right Nipper and RCA logos, swing handle, AM, bat **$25.00**

1-T-4H, vertical, 1960, plastic, eight transistors, upper front round dial, lower thumbwheel on/off/volume knob, center perforated grill area, swing handle, AM, bat **$35.00**

1-T-4J, vertical, 1960, plastic, eight transistors, upper front round dial, lower thumbwheel on/off/volume knob, center perforated grill area, swing handle, AM, bat **$35.00**

1-T-5J, horizontal, 1959, eight transistors, top horizontal slide rule dial, thumbwheel tuning and on/off/volume knobs, front horizontal grill bars, handle, AM, bat **$25.00**

1-TP-1E, vertical, 1961, champagne white, six transistors, upper front round dial knob, right side thumbwheel on/off/volume knob, lattice grill area, AM, bat **$20.00**

1-TP-1HE, vertical, 4x2½x1¼", 1961, Bermuda turquoise/champagne white plastic, six transistors, upper front round dial knob, right side thumbwheel on/off/volume knob, lattice grill area, made in USA, AM, bat $20.00

1-TP-1JE, vertical, 4x2½x1¼", 1961, charcoal/champagne white plastic, six transistors, upper front round dial knob, right side thumbwheel on/off/volume knob, lattice grill area, AM, bat **$20.00**

1-TP-2E, vertical, 1961, champagne white, six transistors, upper front round dial with top thumbwheel tuning, lower perforated grill area with lower right Nipper logo, AM, bat **$25.00**

1-TP-2J, vertical, 1961, charcoal, six transistors, upper front round dial with top thumbwheel tuning, lower perforated grill area with lower right Nipper logo, AM, bat **$25.00**

3RG14, vertical, 1962, six transistors, upper front round dial overlaps large lower checkered grill area, swing handle, AM, bat **$20.00**

3RG81 "Globe Trotter," horizontal, 1963, eight transistors, upper front horizontal slide rule dial, large lower grill area with knob, handle, AM, bat **$15.00**

3RH10, vertical, 4x2½x1⅛", 1960, plastic, upper front round dial knob, right side thumbwheel on/off/volume knob, lattice grill area, made in USA, AM, bat $20.00

3RH21G, vertical, 4x2½x1⅛", 1960, plastic, upper front round dial knob, right side thumbwheel on/off/volume knob, lattice grill area, made in USA, AM, bat $20.00

4RG26, horizontal, 4½x6¾x1½", plastic, upper right front round window dial with right side thumbwheel tuning, lower right thumbwheel on/off/volume knob, left metal perforated grill area, pull-up handle, AM, bat $20.00

4RG51, vertical, 6½x4⅛x1½", 1963, plastic, eight transistors, upper right front turquoise window dial with thumbwheel tuning, thumbwheel on/off/volume knob, left metal perforated grill area with horizontal lines, pull-up handle, AM, bat $25.00

4RG56, vertical, 6½x4⅛x1½", 1963, plastic, eight transistors, upper right front turquoise window dial with thumbwheel tuning, thumbwheel on/off/volume knob, left metal perforated grill area with horizontal lines, pull-up handle, AM, bat **$25.00**

4RG61, horizontal, 1963, plastic, eight transistors, upper right front window dial with thumbwheel tuning, lower right front thumbwheel on/off/volume knob, large left metal perforated grill area with lower left logo, handle, AM, bat **$25.00**

4RM41, horizontal, 1964, 12 transistors, three left front vertical dial scales, four knobs, right perforated grill area, telescoping antenna, AM, FM, SW, bat **$15.00**

7-BT-9J, horizontal, 3½x5¾x1½", 1958, plastic, RCA's first transistor radio, six transistors, diagonally divided front with right front round dial overlapping perforated grill area, AM, bat **$200.00**

7-BT-10K, horizontal, 6½x10x3¾", 1955, leather, seven transistors, upper front horizontal dial, right and left side knobs, horizontal metal grill bars, leather handle, AM, bat $35.00

8-BT-7LE, horizontal, 3¼x5½x1½", 1957, plastic, four transistors, right front round dial knob over horizontal wrap-around grill bars, lower right thumbwheel on/off/volume knob, AM, bat $65.00

8-BT-8FE, horizontal, 3¼x5½x1½", 1957, plastic, four transistors, right front round dial knob over horizontal wrap-around grill bars, lower right thumbwheel on/off/volume knob, AM, bat $65.00

8-BT-10K, horizontal, 1957, leather, seven transistors, upper front horizontal dial, right and left side knobs, horizontal metal grill bars, leather handle, AM, bat **$45.00**

9-BT-9E, horizontal, 1957, white, six transistors, right front round dial knob overlaps horizontal grill bars, top right thumbwheel on/off/volume knob, AM, bat................ **$40.00**

9-BT-9H, horizontal, 1957, green, six transistors, right front round dial knob overlaps horizontal grill bars, top right thumbwheel on/off/volume knob, AM, bat................ **$40.00**

9-BT-9J, horizontal, 1957, gray, six transistors, right front round dial knob overlaps horizontal grill bars, top right thumbwheel on/off/volume knob, AM, bat................ **$40.00**

RFG25E, horizontal, 5⅝x7¼x2½", 1964, leatherette/plastic, eight transistors, two upper right front

knobs – right tuning, left on/off/volume – lower textured grill area, handle, AM, bat $10.00

RFG35, horizontal, 5½x8½x2¾", 1964, leather, eight transistors, right front vertical slide rule dial and two knobs, large left textured grill area, leather handle, AM, bat **$15.00**

RGG25B, horizontal, 5½x7½x2½", tan/ivory leatherette/plastic, eight transistors, two upper right front knobs – right tuning, left on/off/volume – lower textured grill area, handle, AM, bat $15.00

RGG25E, horizontal, 5½x7½x2½", black leatherette/plastic, eight transistors, two upper right front knobs – right tuning, left on/off/volume – lower textured grill area, handle, AM, bat **$15.00**

RGG25G, horizontal, 5½x7½x2½", olive/black leatherette/plastic, eight transistors, two upper right front knobs – right tuning, left on/off/volume – lower textured grill area, handle, AM, bat **$15.00**

RGG29E "Globetrotter," horizontal, 1965, eight transistors, upper right front dial knob, lower on/off/volume knob, large left textured grill area, handle, AM, bat **$15.00**

RJG 15E, vertical, 4x2¾x1¼", black/white plastic, upper right front square window dial with right side thumbwheel tuning, right side thumbwheel on/off/volume knob, large front metal perforated grill area, AM, bat **$20.00**

RJG 15G, vertical, 4x2¾x1¼", olive/ivory plastic, upper right front square window dial with right side thumbwheel tuning, right side thumbwheel on/off/volume knob, large front metal perforated grill area, AM, bat **$20.00**

RJG 15Y, (bottom left) vertical, 4x 2¾ x1¼", white/blue plastic, upper right front square window dial with right side thumbwheel tuning, right side thumbwheel on/off/volume knob, large front metal perforated grill area, AM, bat $20.00

T-1EH, vertical, 7x4¼x2⅛", 1959, Bermuda turquoise and champagne white plastic, six transistors, upper front round dial knob, lower right side on/off/volume knob, horizontal grill bars, swing handle, AM, bat $25.00

T-1EN, vertical, 7x4¼x2⅛", 1959, Monterey red and champagne white plastic, six transistors, upper front round dial knob, lower right side on/off/volume knob, horizontal grill bars, swing handle, AM, bat **$25.00**

T-1JE, vertical, 7x4¼x2⅛", 1959, charcoal and champagne white plastic, six transistors, upper front round dial knob, lower right side on/off/volume knob, horizontal grill bars, swing handle, AM, bat $25.00

T-2K, (bottom left) vertical, 1960, leather, six transistors, upper front round dial knob, lower right side on/off/volume knob, horizontal grill bars, handle, AM, bat.... $20.00

TX-1JE, horizontal/table, 1960, six transistors, center front vertical dial, right and left grill areas with horizontal bars, upper right thumbwheel on/off/volume knob, dual speakers, AM, bat **$15.00**

Realistic

90L611, horizontal, 1962, leather, 10 transistors, upper right front window dial with right side thumbwheel tuning, upper left thumbwheel on/off/volume knob, lower grill area with circular cut-outs and lower left logo, handle, AM, bat **$30.00**

90L665, vertical, 1962, six transistors, upper front window dial with right thumbwheel tuning, left thumbwheel on/off/volume knob, lower perforated grill area with center logo, AM, bat **$30.00**

90L696, horizontal, 1961, leather, eight transistors, off-center front window dial with large right thumbwheel tuning knob, top left thumbwheel on/off/volume knob, square grill cut-outs, handle, AM, bat **$30.00**

90LX661, horizontal, 1961, nine transistors, upper front horizontal three-band slide rule dial with thumbwheel tuning, lower perforated grill area, right sensitivity/

tone/band switches, telescoping antenna, AM, SW, LW, bat **$35.00**

1283K, vertical, 1965, 10 transistors, upper front horizontal two-band slide rule dial with thumbwheel tuning, lower horizontal grill bars, telescoping antenna, strap, AM, FM, bat **$15.00**

1284K, horizontal, 1965, 10 transistors, right front dial with thumbwheel tuning, lower on/off/volume knob, large left grill area, two switches, telescoping antenna, AM, FM, bat **$15.00**

"Hi-Fiver," horizontal, 3x5¼x1⅝", plastic, right front thumbwheel dial, top thumbwheel on/off/volume knob, left metal perforated grill area with upper left logo, made in Japan, AM, bat $40.00

Realtone

TR-555, vertical, 1960, four transistors, upper front window dial with right side thumbwheel tuning, right side thumbwheel on/off/volume knob, metal grill area with horizontal slots, AM, bat......................... **$50.00**

TR-561 "Venus," vertical, 1962, four transistors, upper right front round window dial with right side thumbwheel tuning, left side thumbwheel on/off/volume knob, circular perforated grill area, AM, bat......... **$45.00**

TR-801 "Electra," vertical, 4x2½ x1", 1960, plastic, six transistors, upper front window dial with right side thumbwheel tuning, right side thumbwheel on/off/volume knob, round metal perforated grill area, rear fold-out stand, AM, bat $50.00

TR-803, vertical, 3⅜x2½x1", plastic with metal side trim, upper front window dial with right front thumbwheel tuning, left front thumbwheel on/off/volume knob, round metal perforated grill area, made in Japan, AM, bat **$55.00**

TR-804-2, vertical, 1962, six transistors, upper front window dial with right front thumbwheel tuning, left front thumbwheel on/off/volume knob, lower perforated grill area, AM, bat **$30.00**

TR-806-1 "Ultima," horizontal, 1962, six transistors, upper right front window dial with right and left thumbwheel knobs, large lower perforated grill area, AM, bat **$40.00**

TR-806B, horizontal, 1963, six transistors, upper right front window dial with right and left thumbwheel knobs, large lower perforated grill area, AM, bat **$40.00**

TR-861 "Pioneer," vertical, 3⅞x2⅝x 1¼", 1962, plastic, six transistors, upper right front round window dial with right side thumbwheel tuning, left side thumbwheel on/off/volume knob, round metal perforated grill area, AM, bat.......................... **$45.00**

TR-861-1, vertical, 1962, six transistors, upper right front round window dial with right side thumbwheel tuning, left side thumbwheel on/off/ volume knob, metal perforated grill area, AM, bat.......................... **$45.00**

TR-870 "Satellite," (top right) vertical, 4⅝x2¾x1¼", plastic, six transistors, top raised see-through two-band dial with right side thumbwheel tuning, right side thumbwheel on/off/ volume knob, left side BC/SW switch, vertical metal grill bars, rear fold-out stand, made in Japan, AM, SW, bat $65.00

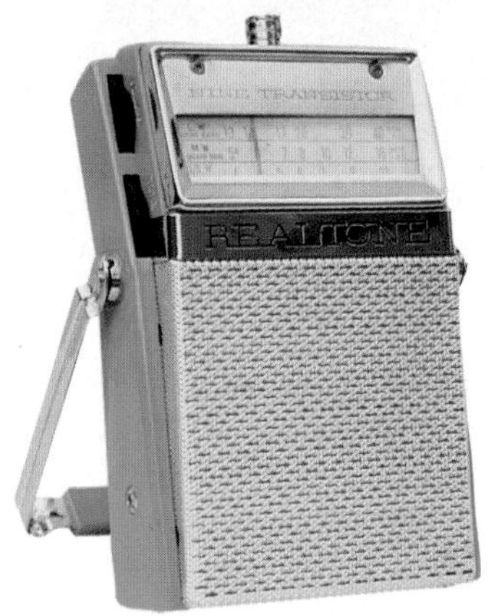

TR-970, (above) vertical, 6x4x2", 1963, plastic, nine transistors, case arches backwards, upper front horizontal three-band slide rule dial with right side thumbwheel tuning, right side LW/MW/SW switch, left side thumb-wheel volume and tone knobs, lower metal perforated grill area, telescoping antenna, swing handle, made in Japan, SW, LW, MW, bat $45.00

TR-1030, horizontal, 1963, leather, 10 transistors, right front window dial, upper left on/off/volume knob, horizontal grill bars, leather handle, AM, bat **$15.00**

TR-1053, vertical, 1964, 10 transistors, upper front horizontal slide rule dial with thumbwheel tuning, lower vertical grill bars, AM, bat **$15.00**

TR-1055 "Duo-Fi," horizontal, 1964, leather, 10 transistors, upper right front window dial, two knobs, left grill area with oblong cut-outs, H/L switch, leather handle, AM, bat **$15.00**

TR-1057, horizontal, 1963, 10 transistors, right front dial with thumbwheel tuning, large left perforated grill area, AM, bat **$20.00**

TR-1069, vertical, 4½x3x1½", 1965, 10 transistors, upper right front window dial with right side thumbwheel tuning, lower perforated grill area, AM, bat **$10.00**

TR-1088 "Comet," vertical, 4¼x2¾x 1¼", 1962, plastic, eight transistors, upper front round window dial inside "figure eight" trim, right side thumbwheel tuning & on/off/volume knobs, lower round metal perforated grill area, rear fold-out stand, made in Japan, AM, bat **$65.00**

TR-1256 "Duo-Fi," horizontal, 1963, leather, 12 transistors, upper right front window dial, upper left on/off/volume knob, horizontal grill bars, leather handle, AM, bat **$10.00**

TR-1618, horizontal, 1963, six transistors, off-center vertical slide rule dial divides large perforated grill area, right side thumbwheel tuning and on/off/volume knobs, AM, bat **$20.00**

TR-1623, horizontal/watch radio, 1963, six transistors, upper right front window dial with right and left thumbwheel knobs, left round watch face and auto thumbwheel knob, lower metal perforated grill area, AM, bat **$70.00**

TR-1623B, horizontal/watch radio, 1963, six transistors, upper right front window dial with right and left thumbwheel knobs, left round watch face and auto thumbwheel knob, lower metal perforated grill area, AM, bat $70.00

TR-1628, vertical, 1963, six transistors, upper right front window dial with thumbwheel tuning, lower perforated grill area, AM, bat **$15.00**

TR-1645, vertical, 1963, plastic, six transistors, step-back top, upper left front window dial over large grill area with horizontal bars, upper left thumbwheel tuning, right side thumbwheel on/off/volume knob, AM, bat $15.00

TR-1660, vertical, 1964, plastic, six transistors, upper right front window dial with right side thumbwheel tuning, large grill area with horizontal bars, AM, bat $15.00

TR-1675, vertical, 1965, six transistors, upper right front window dial with right side thumbwheel tuning, large lower grill area with vertical bars, AM, bat $10.00

TR-1758, vertical, 1963, seven transistors, upper right front window dial with right side thumbwheel tuning, lower perforated grill area, AM, bat $15.00

TR-1820, vertical, 4x2½x1¼", 1962, plastic, eight transistors, upper front horizontal slide rule dial with

thumbwheel tuning, lower metal perforated grill area, AM, bat $25.00

TR-1843, vertical, 4¼x2⅝x1¼", 1963, plastic, eight transistors, upper front window dial with right side thumbwheel tuning, right side thumbwheel on/off/volume knob, large metal perforated grill area, made in Japan, AM, bat $40.00

TR-1844, horizontal, 1963, eight transistors, right front dial and on/off/volume knobs, left grill area with circular cut-outs, leather handle, AM, bat **$10.00**

TR-1859, vertical, 1964, eight transistors, upper front window dial with right side thumbwheel tuning, large lower grill area with horizontal bars, AM, bat **$10.00**

TR-1871, vertical, 1965, eight transistors, upper right front window dial with right side thumbwheel tuning, large lower grill area with vertical bars, AM, bat **$10.00**

TR-1887, vertical, 1965, eight transistors, upper right front round window dial with thumbwheel tuning, large lower grill area with vertical bars, AM, bat **$10.00**

TR-1929, horizontal, 1963, nine transistors, right front dial with right side thumbwheel tuning, right side thumbwheel on/off/volume knob, large left perforated grill area, AM, bat **$20.00**

TR-1946, vertical, 1963, nine transistors, upper right front window dial with right side thumbwheel tuning, large lower grill area with vertical bars, AM, bat **$15.00**

TR-1948, vertical, 1964, nine transistors, upper right front window dial with right side thumbwheel tuning, large lower grill area with horizontal bars, AM, bat **$15.00**

TR-2001, horizontal, 1962, 14 transistors, upper front horizontal two-band slide rule dial with thumbwheel tuning, large lower perforated grill area, two telescoping antennas in handle, AM, FM, bat **$25.00**

TR-2021, horizontal, 1963, ten transistors, upper front horizontal two-band slide rule dial, lower grill area, four knobs, two telescoping antennas, AM, FM, bat **$25.00**

TR-2051, horizontal, 1963, 10 transistors, right front vertical two-band slide rule dial, right and left front thumbwheel knobs, large grill area with horizontal bars, telescoping antenna, AM, FM, bat $20.00

TR-2076, horizontal, 1965, 10 transistors, upper front horizontal two-band slide rule dial, lower grill area with vertical bars and lower right FM/AM switch, telescoping antenna, AM, FM, bat **$15.00**

TR-2864, horizontal, 1964, eight transistors, upper front horizontal two-band slide rule dial, large lower grill area with vertical bars, MW/SW switch, telescoping antenna, AM, SW, bat **$15.00**

TR-2884, horizontal, 1965, eight transistors, right front two-band window dial with top thumbwheel tuning and on/off/volume knobs, left grill area with horizontal bars, AM, FM, bat **$15.00**

TR-2925, horizontal, 1963, nine transistors, upper front horizontal two-band slide rule dial, thumbwheel tuning and on/off/volume knobs, round perforated grill area, telescoping antenna, AM, SW, bat **$25.00**

TR-3047, horizontal, 1963, 10 transistors, three right front vertical dial scales, large left grill area with vertical bars, telescoping antenna, AM, SW, LW, bat **$15.00**

TR-3422, horizontal, 1963, 14 transistors, upper front horizontal three-band slide rule dial with thumbwheel tuning, large lower perforated grill area, two telescoping antennas in handle, AM, FM, SW, bat **$25.00**

TR-4016, horizontal, 1963, 10 transistors, upper front horizontal four-band slide rule dial, large lower perforated grill area, thumbwheel knobs, telescoping antenna, handle, AM, 2SW, LW, bat **$25.00**

TR-8611 "Constellation," (bottom left) vertical, 3$\frac{3}{4}$x2$\frac{1}{2}$x1", 1963, plastic, six transistors, upper right front round window dial with right side thumbwheel tuning, left side thumbwheel on/off/volume knob, metal perforated grill area, rear fold-out stand, AM, bat $45.00

Regency

TR-1, vertical, 5x3x1$\frac{1}{4}$", 1954, available in Mandarin red, cloud gray, ivory, black, jade green, and mottled mahogany plastic, four transistors, the first transistor radio, large upper right front round brass dial knob, upper left thumbwheel on/off/volume knob, lower perforated grill area, AM, bat.

Mandarin red $350.00
cloud gray $350.00
ivory $300.00
black $250.00
jade green $600.00
mottled mahogany $550.00

TR-1G, vertical, 5x3x1¼", 1958, available in coral, yellow, black, turquoise, or ivory plastic, four transistors, large upper right front round dial knob with inner concentric circles, upper left thumbwheel on/off/volume knob, lower perforated grill area, AM, bat.
coral **$275.00**
yellow **$250.00**
black **$150.00**
turquoise **$250.00**
ivory **$175.00**

TR-4, vertical, 5x3x1¼", 1957, available in ebony, red, or ivory plastic, four transistors, large upper right front round dial knob with inner concentric circles, upper left thumbwheel on/off/volume knob, lower perforated grill area, AM, bat.
ebony **$100.00**
red **$200.00**
ivory **$150.00**

TR-5, horizontal, 3½x6½x2", 1958, leather, five transistors, right front round brass dial knob, lower on/off/volume knob, left grill area with oblong cut-outs and lower left logo, leather handle, AM, bat **$45.00**

TR-5A, 3½x6½x2", 1958, leather, five transistors, right front round brass dial knob, lower on/off/volume knob, left grill area with oblong cut-outs and lower left logo, leather handle, AM, bat **$45.00**

TR-5B, 3½x6½x2", 1958, leather, five transistors, right front round brass dial knob, lower on/off/volume knob, left grill area with oblong cut-outs and lower left logo, leather handle, AM, bat **$45.00**

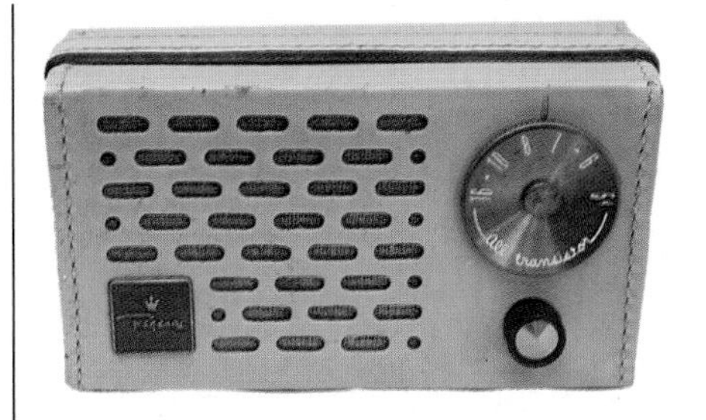

TR-5C, horizontal, 3½x6½x2", 1958, Briarwood, California saddle, or champagne leather, five transistors, right front round brass dial knob, lower on/off/volume knob, left grill area with oblong cut-outs and lower left logo, leather handle, AM, bat $45.00

TR-6, horizontal, 5¼x7⅛x3⅛", 1957, leather, six transistors, right side dial knob, left side on/off/volume knob, front grill area with circular cut-outs, leather handle, AM, bat $55.00

TR-7, vertical, 5¾x3½x2", 1958, available in black or ivory plastic, seven transistors, upper left front round dial knob, upper right front on/off/volume knob, lower patterned grill area, swing handle, AM, bat **$70.00**

TR-11, vertical, 5x3x1¼", 1959, available in red, white, or ebony plastic, four transistors, upper front round dial knob over horizontal grill bars, AM, bat **$65.00**

TR-22, horizontal, 1959, leather, four transistors, right side dial knob, left side on/off/volume knob, large front grill area, leather handle, AM, bat **$35.00**

TR-99 "World Wide," vertical, 5⅞x 3⅝x2⅜", 1960, available in white or ebony plastic, seven transistors, upper left front round dial, upper right on/off/volume knob, lower plastic perforated random-patterned grill area with lower right logo, swing handle, AM, bat $90.00

XR-2A, vertical, 3x2⅛x1⅛", 1958, red plastic, two transistors, large front round dial, top on/off/switch, no

speaker, earphone is wired in, no volume control, AM, bat $100.00

Rhapsody

FA-101, horizontal, 1964, 10 transistors, two upper front horizontal slide rule dials – one AM, one FM – large lower perforated grill area, telescoping antenna, AM, FM, bat **$20.00**

TR8A7 "High Fidelity," vertical, 1963, eight transistors, upper left front round dial knob over large perforated grill area, right side thumbwheel on/off/volume knob, AM, bat **$25.00**

Riviera

RV62, vertical, 1962, six transistors, upper right front window dial with right side thumbwheel tuning, top left thumbwheel on/off/volume knob, lower perforated grill area, AM, bat **$25.00**

Robin

TR-605, vertical, 1960, six transistors, upper left front window dial with top left thumbwheel tuning, upper right thumbwheel on/off/volume knob, lower lattice grill area, AM, bat $35.00

Roland

4TR, horizontal, 1959, four transistors, top left dial knob, top right on/off/volume knob, large lower grill area with horizontal bars and lower right logo, handle, AM, bat ... **$35.00**

6TR "All Transistor 66," vertical, 6¼x 4¾x2⅝", 1957, leather, six transistors, top right dial knob, top left on/off/volume knob, front grill area with rectangular cut-outs, leather strap, AM, bat $45.00

51-481 (TW4) "Bi-Fidelity," horizontal, 1960, five transistors, upper front thumbwheel dial knob, large lower grill area with horizontal bars, logo and on/off/volume knob, swing handle, AM, bat **$25.00**

61-482 (5TR), horizontal, 5x6⅜x2⅝", 1959, available in white, ebony, or red plastic, six transistors, top left dial knob, top right on/off/volume knob, large lower grill area with horizontal bars and lower right logo, handle, AM, bat **$35.00**

71-483 (TW6) "Bi-Fidelity," horizontal, 1959, seven transistors, upper front thumbwheel dial, large lower checkered grill area with on/off/volume knob and logo, right and left round grills with horizontal bars, swing handle, AM, bat **$50.00**

71-486 (7TW), horizontal, 9¼x11½x 4⅛", 1959, available in walnut, mahogany, blond, cherry, or ebony wood, seven transistors, top raised slide rule dial, right and left knobs, large lower perforated grill area with twin speakers, wire stand, handle, AM, bat **$35.00**

TC-10, horizontal/clock radio, 1960, seven transistors, lower right front half-round dial, upper lattice grill area, left alarm clock face, handle, AM, bat **$30.00**

TR8 "Bi-Fidelity," vertical, 1960, available in black or tan, seven transistors, top dial and on/off/volume knobs, large front perforated grill area, strap, AM, bat **$35.00**

Ronith

101, vertical, 4¼x2¾x1⅜", plastic, 10 transistors, upper front round dial knob, right side thumbwheel on/off/volume knob, lower lattice grill area, made in Hong Kong, AM, bat **$20.00**

Roscon

8TS-33 "Super," vertical, 1962, eight transistors, upper right front round dial knob, upper left front on/off/volume window, lower perforated grill area, AM, bat **$30.00**

KR-6TS-40, vertical, 1962, six transistors, upper left front window dial with left side thumbwheel tuning, upper right front on/off/volume window with right side thumbwheel knob, lower perforated grill area, AM, bat **$35.00**

Ross

1063, vertical, 4x2⅝x1¼", plastic, solid state, upper right front window dial with right side thumbwheel tuning, left side thumbwheel on/off/volume knob, large metal perforated grill area with lower left logo, vinyl strap, made in Hong Kong, AM, bat $15.00

1801 "Magnifique," horizontal, 1964, 13 transistors, upper front horizontal three-band slide rule dial, lower grill area with vertical bars, right side knobs, two telescoping antennas, handle, AM, FM, SW, bat **$30.00**

1904 "Adventurer," horizontal, 1965, upper front horizontal three-band slide rule dial with thumbwheel tuning, large lower grill area with vertical bars, two speakers, telescoping antenna, handle, AM, FM, SW, bat **$20.00**

"Imperial 76," horizontal, 1964, 11 transistors, right front two-band vertical slide rule dial, three knobs, large left perforated grill area, top pushbuttons, telescoping antenna, handle, AM, FM, bat **$20.00**

"Imperial 91," vertical, 1964, nine transistors, upper front round two-band dial with center logo, right side thumbwheel knobs, lower lattice grill area, telescoping antenna, AM, FM, bat **$15.00**

RE-66, horizontal, 1965, six transistors, upper right thumbwheel dial, lower right thumbwheel on/off/volume knob, left perforated grill area with center logo, strap, AM, bat **$25.00**

RE-101 "Dynamic," vertical, 1964, plastic, 10 transistors, upper front round dial knob, right side on/off/volume knob, lower lattice grill area, AM, bat **$20.00**

RE-102, horizontal, leather, 10 transistors, upper right front round dial knob, lower right front thumbwheel on/off/volume knob, left round metal perforated grill area with center Ross logo, AM, bat........... **$20.00**

RE-104 "12," horizontal, 1965, leather, 12 transistors, right front round dial knob, left grill area with horizontal bars, leather handle, AM, bat **$15.00**

RE-104 "14," horizontal, 1965, leather, 14 transistors, right front round dial knob, left grill area with horizontal bars, leather handle, AM, bat **$15.00**

RE-110N "Sportsman," horizontal, 5 3/8x9 1/8x2 1/2", leather, 12 transistors, right front dial with thumbwheel tuning, right side AC/bat switch, upper left front on/off/volume knob over grill area, leather handle, made in Hong Kong, AM, AC/bat $15.00

RE-120, horizontal, 1964, 12 transistors, right front round two-band dial over large perforated grill area, top pushbuttons, telescoping antenna, handle, AM, FM, bat **$25.00**

RE-125, horizontal, 1964, 12 transistors, large right front round two-band dial, right side knob, left grill area with horizontal slots, telescoping antenna, handle, AM, FM, bat **$15.00**

RE-510, horizontal, 1964, 10 transistors, upper front horizontal two-band dial with thumbwheel tuning, large lower perforated grill area, telescoping antenna, handle, AM, FM, bat **$15.00**

RE-700 "Micro," vertical, 1965, seven transistors, right front window dial with right side thumbwheel tuning, lower grill area, AM, bat **$50.00**

RE-714, square, 1964, seven transistors, right side dial and on/off/volume knobs, front round grill area with center "flower" decoration, left side strap, AM, bat **$50.00**

RE777 "Jubilee," vertical, 1964, plastic, seven transistors, upper front round dial knob, right side thumbwheel knob, lower lattice grill area, AM, bat **$15.00**

RE-815 "Micro," square, 2 3/4x2 1/2 x1 1/8", plastic, eight transistors, right side dial and on/off/volume knobs, front metal perforated grill area with lower left logo, left side metal chain and key ring, made in Hong Kong, AM, bat $45.00

RE-900 "Micro," vertical, 1965, nine transistors, left front vertical three-band slide rule dial with top thumbwheel tuning, top thumbwheel on/off/volume knob, large perforated grill area, telescoping antenna, strap, AM, FM, SW, bat **$35.00**

RE-1112 "Jubilee," horizontal, 1964, leather, 12 transistors, upper front round dial knob, right and left round perforated grill areas, right side thumbwheel on/off/volume knob and switch, leather handle, AM, bat **$20.00**

RE-1115 "High Fidelity," horizontal, 1965, leather, 10 transistors, upper right front round dial knob, lower on/off/volume knob, left oval grill area with center logo, handle, AM, bat **$15.00**

RE-1212, vertical, 1965, 12 transistors, upper right front oval window dial with right side thumbwheel tuning, top left thumbwheel on/off/volume knob, oval perforated grill area with center logo, AM, bat **$20.00**

RE-1902 "Magnifique," horizontal, 1964, nine transistors, right front dial with thumbwheel tuning, top left thumbwheel on/off/volume knob, oval perforated grill area with center logo, AM, bat **$20.00**

Sampson

BT65, vertical, 1963, six transistors, upper right front window dial with right side thumbwheel tuning, top left thumbwheel on/off/volume knob, large perforated grill area with lower left "S" logo, AM, bat.... **$15.00**

BT66, vertical, 1963, six transistors, upper right front window dial with right side thumbwheel tuning, top left thumbwheel on/off/volume knob, large perforated grill area, AM, bat $15.00

BT85, horizontal, 1963, eight transistors, right front window dial with right side thumbwheel tuning, large perforated grill area with lower left "S" logo, AM, bat **$20.00**

S-640, vertical, 1962, six transistors, upper right round dial knob, upper left on/off/volume window with top thumbwheel knob, lower perforated grill area, AM, bat **$30.00**

SC4000 "Super Alarm," horizontal/watch radio, 2¾x4¾x1", plastic, six transistors, upper right front window dial with right side thumbwheel tuning, lower right side thumbwheel on/off/volume knob, right metal perforated grill area, left

alarm watch face and "alarm/off/ radio" switch, AM, bat $75.00

Sanyo

AFT-6N, horizontal, 3¾x6¾x2", plastic, nine transistors, top horizontal three-band slide rule dial, large front metal perforated grill area with lower right switch, LW, MW, UKW, bat $30.00

AFT-9S "Transworld," horizontal, 6¾ x10½x2½", plastic, nine transistors, step-back top, upper front horizontal three-band slide rule dial with right thumbwheel tuning knob, left thumbwheel on/off/volume knob, large lower metal perforated grill area with center logo, three top pushbuttons, right side fine tuning knob, left side tone switch, telescoping antenna, handle, AM, FM, SW, bat $25.00

Satelite

"Boy's Radio," vertical, plastic, two transistors, upper right front window dial with thumbwheel tuning, lower metal perforated grill area, made in Japan, AM, bat **$30.00**

Satellite

60N63, vertical, plastic, six transistors, upper right front dial with thumbwheel tuning, left thumbwheel on/off/volume knob, lower perforated grill area, AM, bat $25.00

Saxony

606, vertical, 4¼x2½x1¼", 1963, plastic, six transistors, upper left front round window dial with left side thumbwheel tuning, right side thumbwheel on/off/volume knob, lower metal perforated grill area, made in USA, AM, bat $20.00

Sceptre

"Boy's Radio," vertical, plastic, two transistors, upper front window dial with left side thumbwheel tuning, right side thumbwheel on/off/volume knob, round metal perforated grill area, AM, bat $35.00

Selsi

"Boy's Radio," (bottom left) vertical, 3⅛x2½x1¼", plastic, two transistors, upper right front window dial with right side thumbwheel tuning, left side thumbwheel on/off/volume knob, front round metal perforated grill area, made in Japan, AM, bat $40.00

Seminole

600, vertical, 1962, six transistors, upper front horizontal dial with thumbwheel tuning, lower perforated grill area, AM, bat **$15.00**

605, vertical, 1964, six transistors, upper front off-center round window dial with thumbwheel tuning, lower round perforated grill area, AM, bat **$20.00**

800, horizontal, 1963, eight transistors, right front vertical slide rule dial with right side thumbwheel tuning, lower right front on/off/volume window with right side thumbwheel knob, left perforated grill area, AM, bat.......................... **$20.00**

801, horizontal, 1963, eight transistors, right front panel with round dial knob and decorative "stars," lower right side thumbwheel on/off/volume knob, left perforated grill area, AM, bat **$40.00**

803, horizontal, 1963, eight transistors, step-down top with horizontal two-band dial, right and left thumbwheel knobs, large lower perforated grill area, telescoping antenna, AM, SW, bat **$25.00**

805, vertical, 1964, eight transistors, large recessed and perforated oval front panel with upper "cat's eye" dial, right side thumbwheel knob, AM, bat **$40.00**

806, horizontal, 1964, eight transistors, upper right front dial with upper right side thumbwheel tuning, lower right side thumbwheel on/off/volume knob, left oval perforated grill area, AM, bat **$25.00**

900, horizontal, 1962, nine transistors, upper front horizontal slide rule dial with upper right side thumbwheel tuning, lower right front round on/off/volume window with right side thumbwheel knob, perforated grill area, AM, bat **$30.00**

901, horizontal, 1963, nine transistors, upper front horizontal four-band slide rule dial with thumbwheel tuning, large lower perforated grill area, telescoping antenna, AM, 2SW, LW, bat ... **$25.00**

1000, horizontal, 1962, 10 transistors, upper front horizontal two-band slide rule dial with thumbwheel tuning, lower perforated grill area, BC/SW switch, telescoping antenna, AM, SW, bat **$30.00**

1001, (bottom left) horizontal, 3¼x5½ x1½", 1963, plastic, ten transistors, step-down right side, top slide rule dial with top right thumbwheel tuning, lower right side thumbwheel on/off/volume knob, large metal perforated grill area, AM, bat $40.00

1010, horizontal, 1964, 10 transistors, right front dial over perforated wrap-around panel, right side thumbwheel knobs, AM, bat **$20.00**

1011, vertical, 1964, 10 transistors, large recessed and perforated oval front panel with upper "cat's eye" dial, right side thumbwheel knob, AM, bat **$40.00**

1015, horizontal, 1964, 10 transistors, right front round window dial with right side thumbwheel tuning, right side thumbwheel on/off/volume knob, oval grill area, AM, bat **$30.00**

1020, horizontal, 1964, leather, 10 transistors, right front dial with thumbwheel tuning, large left perforated grill area, handle, AM, bat **$20.00**

1030, horizontal, 1964, 10 transistors, upper front horizontal three-band slide rule dial, large lower grill area, top pushbuttons, telescoping antenna, handle, AM, FM, SW, bat **$20.00**

1100, horizontal, 1962, 11 transistors, upper front horizontal two-band slide rule dial with upper right thumbwheel tuning, left thumbwheel on/off/volume knob, lower perforated grill area, band switch, telescoping antenna, AM, FM, bat **$35.00**

1101, horizontal, 1963, 11 transistors, step-down right side, top slide rule dial with top right thumbwheel tuning, lower right side thumbwheel on/off/volume knob, large perforated grill area, AM, bat **$30.00**

1102, horizontal, 1963, 10 transistors, upper front horizontal two-band slide rule dial, right and left knobs, two lower round perforated grill areas, telescoping antenna, handle, AM, FM, bat **$20.00**

1205, horizontal, 1964, 12 transistors, upper front horizontal two-band slide rule dial, large lower perforated grill area with two knobs, two telescoping antennas, handle, AM, FM, bat **$20.00**

KTR-1022, horizontal, 1964, 10 transistors, upper front horizontal two-band slide rule dial with right side thumbwheel tuning, right side thumbwheel on/off/volume knob, large perforated grill area with band switch and lower left logo, telescoping antenna, AM, FM, bat **$20.00**

TR-221, horizontal, 1963, six transistors, right front panel with round dial knob and decorative "stars," lower right side thumbwheel on/off/volume knob, left perforated grill area, AM, bat **$40.00**

Sentinel

1E500, horizontal, 3¼x5¾x1¾", plastic, upper right front thumbwheel

dial, lower right on/off/volume window with lower right side thumbwheel knob, large metal perforated grill area, AM, bat $150.00

Shalco

M6M, horizontal, 2¾x4¼x1¼", plastic, six transistors, upper right front horseshoe-shaped grill with round dial knob, lower right on/off/volume window with thumbwheel knob, large metal perforated grill area with lower left logo, AM, bat $35.00

Sharp

BP-100, horizontal, 2½x4½x1¼", 1964, six transistors, upper right front window dial with right side thumbwheel tuning, lower right side thumbwheel on/off/volume knob, left grill area with horizontal slots and lower left logo, AM, bat **$15.00**

BP-374, vertical, 1963, seven transistors, upper right front window dial with thumbwheel tuning, large round perforated grill area, AM, bat **$20.00**

BP-460, horizontal, 1963, six transistors, upper right front window dial with right side thumbwheel tuning, lower right side thumbwheel on/off/volume knob, left perforated grill area with lower left logo, AM, bat **$15.00**

BP-485, vertical, 1963, nine transistors, upper front horizontal slide rule dial with right side thumbwheel tuning, left front thumbwheel on/off/volume knob, large lower oval perforated grill area with center logo, swing handle, AM, bat **$20.00**

BX-326, horizontal, 1961, 10 transistors, upper front horizontal two-band slide rule dial with upper right thumbwheel tuning, left thumbwheel on/off/volume knob, band switch, large perforated grill area, AM, SW, bat **$30.00**

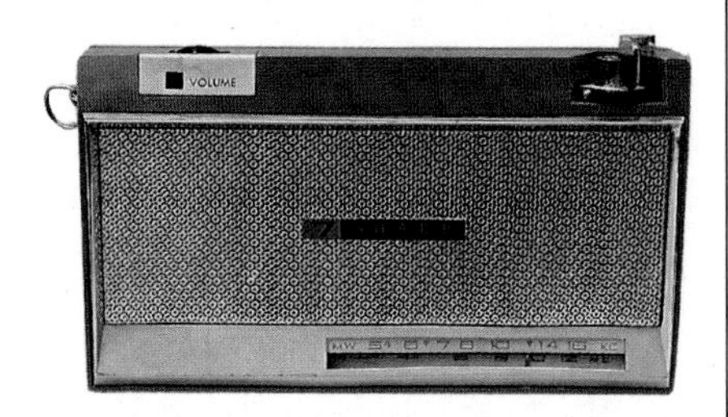

BX-371, horizontal, 3½x6¼x1⅝", plastic, seven transistors, lower right front two-band horizontal slide rule dial with top right thumbwheel tuning and band switch, right side thumbwheel fine tuning knob, upper left volume window with top thumbwheel knob, large metal perforated grill area, rear raised metal plate with model number and address, telescoping antenna, made in Japan, MW, SW, bat $30.00

FW-503, horizontal, 1964, 12 transistors, upper front horizontal four-band slide rule dial, large lower perforated grill area with right knob and lower left logo, telescoping antenna, handle, AM, FM, SW, Marine, bat **$25.00**

FX-109, vertical, 1964, 10 transistors, upper front horizontal two-band slide rule dial with thumbwheel tuning, lower perforated grill area with lower left logo, telescoping antenna, AM, FM, bat **$15.00**

FX-110, vertical, 1965, 10 transistors, upper front horizontal two-band slide rule dial with right side thumbwheel tuning, right side thumbwheel on/off/volume knob, perforated grill area, telescoping antenna, AM, FM, bat **$15.00**

FX-404, horizontal, 6¾x9⅛x2½", plastic, nine transistors, two upper front horizontal dial scales – one AM, one FM – upper left front thumbwheel on/off/volume knob, two pushbuttons, large metal perforated grill area, telescoping antenna in handle, made in Japan, AM, FM, bat **$25.00**

FX-495, horizontal, 1963, 10 transistors, upper front horizontal two-band slide rule dial, lower oval perforated grill area with center logo, telescoping antenna, handle, AM, FM, bat........ **$20.00**

FX-502, horizontal, 1964, 10 transistors, upper front horizontal two-band slide rule dial with right side thumbwheel tuning, large perforated grill area, telescoping antenna, AM, FM, bat.................................. **$15.00**

FY-151M, horizontal, 1965, leather, ten transistors, upper front horizontal three-band slide rule dial, lower left grill area, lower right tuning knob and thumbwheel on/off/volume knob, right side switch, telescoping antenna, leather handle, AM, FM, Marine, bat **$20.00**

FYS-151, horizontal, 1965, leather, ten transistors, upper front horizontal three-band slide rule dial, lower left grill area, lower right tuning knob and thumbwheel on/off/volume knob, right side switch, telescoping antenna, leather handle, AM, FM, SW, bat.................. **$20.00**

TR-173 "Collie," horizontal, 3¼x 5¾x1½", plastic, six transistors, upper right front round dial knob overlaps large metal grill area with vertical slots, right side thumbwheel on/off/volume knob, AM, bat.... $30.00

TR-182, horizontal, 1959, six transistors, right front window dial over large perforated grill area with upper left logo, two right side thumbwheel knobs – one tuning, one on/off/volume – AM, bat **$50.00**

TR-202, horizontal, 2½x4¼x1¼", plastic, right front window dial with thumbwheel tuning, left metal perforated grill area with left logo, AM, bat....................................... $30.00

TR-203, horizontal, 1962, eight transistors, upper front horizontal two-band slide rule dial with upper right thumbwheel tuning, upper left thumbwheel on/off/volume knob, lower perforated grill area, telescoping antenna, AM, SW, bat...... **$30.00**

Shaw

6TR6, vertical, 1965, six transistors, upper right front window dial with right side thumbwheel tuning, top left thumbwheel on/off/volume knob, lower checkered grill area, AM, bat **$15.00**

8TR8, horizontal, 1965, eight transistors, upper right front window dial with thumbwheel tuning, large oval perforated grill area, AM, bat **$25.00**

9FM190, horizontal, 1965, nine transistors, upper front horizontal two-band slide rule dial, large lower perforated grill area with lower right logo, telescoping antenna, handle, AM, FM, bat **$15.00**

10TR10, horizontal, 1965, 10 transistors, upper right front window dial with right side thumbwheel tuning, top thumbwheel on/off/volume knob, large oval perforated grill area, AM, bat **$25.00**

Silvertone

19, horizontal/table, 1964, six transistors, top raised horizontal slide rule dial, three knobs, large lower grill area, AM, bat **$20.00**

20, horizontal/table, 1964, six transistors, top raised horizontal slide rule dial, three knobs, large lower grill area, AM, bat **$20.00**

21, horizontal/table, 1964, six transistors, top raised horizontal slide rule dial, three knobs, large lower grill area, AM, bat **$20.00**

22, horizontal/table, 1964, six transistors, top raised horizontal slide rule dial, three knobs, large lower grill area, AM, bat **$20.00**

206, vertical, 1960, four transistors, upper right front quarter-round window dial with right side thumbwheel tuning, upper left thumbwheel on/off/volume knob, lower V-shaped grill bars, AM, bat **$30.00**

208 "500," horizontal, 1960, five transistors, upper front round dial knob overlaps left horizontal grill bars, AM, bat **$15.00**

211, vertical, 6¼x3¼x1½", 1959, black plastic, six transistors, upper right front double window dial – upper window shows dial numbers, lower window shows CD marks – right side thumbwheel dial knob, left side thumbwheel on/off/volume knob, lower metal perforated grill area, swing handle, AM, bat $30.00

212, vertical, 6¼x3¼x1½", 1959, coral plastic, six transistors, upper right front double window dial – upper window shows dial numbers, lower window shows CD marks – right side thumbwheel dial knob, left side thumbwheel on/off/volume knob, lower metal perforated grill area, swing handle, AM, bat **$30.00**

213, vertical, 6¼x3¼x1½", 1959, ice blue plastic, six transistors, upper right front double window dial – upper window shows dial numbers, lower window shows CD marks – right side thumbwheel dial knob, left side thumbwheel on/off/volume knob, lower metal perforated grill area, swing handle, AM, bat $30.00

214, horizontal, 1960, six transistors, upper front window dial with top thumbwheel tuning, upper right thumbwheel on/off/volume knob, left grill area with horizontal bars, handle, AM, bat **$30.00**

217 "600," horizontal, 5¼x8½x3⅛", 1960, leather, six transistors, right side dial knob, left side on/off/volume knob, front grill area with rectangular cut-outs, leather handle, AM, bat **$15.00**

220 "700," horizontal, 1960, leather, seven transistors, right side dial knob, left side on/off/volume knob, front lattice grill area, leather handle, AM, bat **$15.00**

222 "800," horizontal, 1960, leather, eight transistors, upper front horizontal four-band slide rule dial, lower lattice grill area, telescoping antenna, leather handle, bat **$20.00**

1016, horizontal/table, 1964, ivory, six transistors, wedge-shaped case, upper front horizontal dial, two knobs, large lower grill area, AM, bat **$15.00**

1017, horizontal/table, 1964, ivory front/ice blue back, six transistors, wedge-shaped case, upper front horizontal dial, two knobs, large lower grill area, AM, bat **$15.00**

1018, horizontal/table, 1964, ivory front/coral back, six transistors, wedge-shaped case, upper front horizontal dial, two knobs, large lower grill area, AM, bat **$15.00**

1019 "Medalist," horizontal/table, 1961, ivory front/brown back, seven transistors, three upper front knobs – volume, tone and tuning – large lower grill area, feet, AM, bat **$15.00**

1044, horizontal/clock radio, 1961, five transistors, upper right front round dial knob, upper left alarm clock face, lower grill area with horizontal bars, feet, AM, bat **$15.00**

1201, vertical, 1961, four transistors, upper front thumbwheel dial, right side thumbwheel on/off/volume knob, lower horizontal wrap-around grill bars, AM, bat **$25.00**

1202, vertical, 4x2½x1¼", 1961, black plastic, six transistors, upper right side thumbwheel dial knob, left side thumbwheel on/off/volume knob, metal perforated grill area, AM, bat **$25.00**

1203, vertical, 4x2½x1¼", 1961, mint green plastic, six transistors, upper right side thumbwheel dial knob, left side thumbwheel on/off/volume knob, metal perforated grill area, AM, bat $25.00

1205, vertical, 4¾x3x1½", 1961, black plastic, six transistors, upper right front half-round window dial with right side thumbwheel tuning, upper left on/off/volume window with left side thumbwheel knob, metal perforated grill area, swing handle, AM, bat **$25.00**

1206, vertical, 4¾x3x1½", 1961, ice blue plastic, six transistors, upper right front half-round window dial with right side thumbwheel tuning, upper left on/off/volume window with left side thumbwheel knob, metal perforated grill area, swing handle, AM, bat **$25.00**

1208 "Medalist," horizontal, 3½x7x 1⅝", 1961, black plastic, seven transistors, upper front window dial with top thumbwheel tuning, upper right thumbwheel on/off/volume knob, large metal perforated wrap-around grill area, swing handle, made in USA, AM, bat **$25.00**

1209 "Medalist," horizontal, 3½x7x 1⅝", 1961, ice blue plastic, seven transistors, upper front window dial with top thumbwheel tuning, upper right thumbwheel on/off/volume knob, large metal perforated wrap-around grill area, swing handle, made in USA, AM, bat $25.00

1215 "600," horizontal, 1961, brown leather, six transistors, right side dial knob, left side on/off/volume knob, front perforated grill area, leather handle, AM, bat **$20.00**

1216 "600," horizontal, 1961, gray leather, six transistors, right side dial knob, left side on/off/volume knob, front perforated grill area, leather handle, AM, bat **$20.00**

1217 "700," horizontal, 1961, leather, seven transistors, right side dial knob, left side on/off/volume knob, large front lattice grill area, leather handle, AM, bat **$20.00**

2016 "Medalist," horizontal/table, 1962, eight transistors, three upper front knobs – volume, tone, and tuning – large lower grill area, feet, AM, bat **$15.00**

2201, vertical, 1962, five transistors, upper right front wedge-shaped window dial with right side thumbwheel tuning, lower textured grill area, AM, bat **$15.00**

2205, vertical, 1962, black, six transistors, upper right front wedge-shaped window dial with right side thumbwheel tuning, left side thumbwheel on/off/volume knob, lower "woven" grill area, AM, bat ... $15.00

2206, vertical, 1962, gold, six transistors, upper right front wedge-shaped window dial with right side thumbwheel tuning, left side thumbwheel on/off/volume knob, lower "woven" grill area, AM, bat **$15.00**

2207, vertical, 1962, ice blue, six transistors, upper right front wedge-shaped window dial with right side thumbwheel tuning, left side thumbwheel on/off/volume knob, lower "woven" grill area, AM, bat **$15.00**

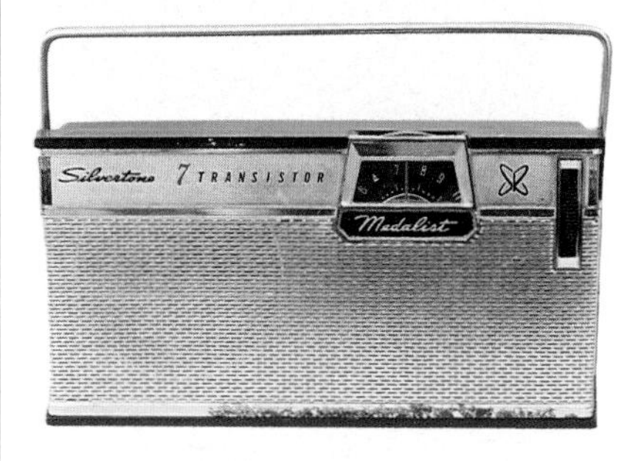

2208 "Medalist," horizontal, 3½x7x 1⅝", 1962, black plastic, seven transistors, upper front window dial with top thumbwheel tuning, upper right thumbwheel on/off/volume knob, large metal perforated wrap-around grill area, swing handle, made in USA, AM, bat $25.00

2209 "Medalist," horizontal, 3½x7x 1⅝", 1962, ice blue plastic, seven transistors, upper front window dial with top thumbwheel tuning, upper right thumbwheel on/off/volume knob, large metal perforated wrap-around grill area, swing handle, made in USA, AM, bat **$25.00**

2212 "500," horizontal, 1962, brown, five transistors, off-center front dial knob overlaps horizontal grill bars, AM, bat **$15.00**

2213 "500," horizontal, 1962, blue, five transistors, off-center front dial knob overlaps horizontal grill bars, AM, bat **$15.00**

2214 "Medalist," horizontal, 1962, eight transistors, large off-center round two-band dial overlaps perforated grill area, two right knobs, telescoping antenna, handle, AM, Marine, bat **$25.00**

2215 "600," horizontal, 5¼x7¾x3¼", 1961, brown leather, six transistors, right side dial knob, left side on/off/ volume knob, large front metal perforated grill area with lower right logo, leather handle, made in USA, AM, bat **$20.00**

2216 "600," horizontal, 5¼x7¾x3¼", 1961, gray leather, six transistors, right side dial knob, left side on/ off/volume knob, large front metal perforated grill area with lower right logo, leather handle, made in USA, AM, bat $20.00

2222 "800," horizontal, 6¼x9¾x3¾", 1962, leather, eight transistors, right side dial knob, left side on/off/volume knob, large front plastic lattice grill area with round logo, leather handle, AM, bat $20.00

2223 "800," horizontal, 6¼x9¾x3¾", 1962, gray leather, eight transistors, right side dial knob, left side on/off/ volume knob, large front plastic lattice grill area with round logo, leather handle, AM, bat **$20.00**

2224, horizontal, 1962, eight transistors, three right front vertical slide rule dial scales, three knobs, left perforated grill area, telescoping antenna, handle, AM, 2SW, bat **$15.00**

2226, horizontal, 1962, 10 transistors, upper front horizontal two-band slide rule dial, two knobs, FM/AM switch, large lower perforated grill area, telescoping antenna, handle, AM, FM, bat **$20.00**

3205, vertical, 3⅜x2½x1⅛", black plastic, six transistors, upper right front window dial with right side thumbwheel tuning, left side thumbwheel on/off/volume knob, metal "woven" grill area, AM, bat **$15.00**

3206, vertical, 3³/₈x2¹/₂x1¹/₈", gold plastic, six transistors, upper right front window dial with right side thumbwheel tuning, left side thumbwheel on/off/volume knob, metal "woven" grill area, AM, bat ... $15.00

3207, vertical, 3⅜x2½x1⅛", ice blue plastic, six transistors, upper right front window dial with right side thumbwheel tuning, left side thumbwheel on/off/volume knob, metal "woven" grill area, AM, bat **$15.00**

3229, horizontal, 1963, 13 transistors, upper front horizontal two-band slide rule dial, two knobs, FM/AFC/AM switch, large lower perforated grill area, telescoping antenna, handle, feet, AM, FM, bat **$20.00**

4211, horizontal, 3¼x7¼x1¾", 1963, eight transistors, metal flip-up front, inner left horizontal slide rule dial with right thumbwheel tuning and on/off/volume knobs, large lower

metal perforated grill area, AM, bat $30.00

5201, vertical, 1964, six transistors, upper right front round dial knob, large textured grill area, AM, bat **$10.00**

5201-A, vertical, 1964, six transistors, upper right front round dial knob, large textured grill area, AM, bat **$10.00**

5202, vertical, 4x2½x1¼", 1965, black plastic, seven transistors, upper right front round dial knob overlaps left textured grill area, left side thumbwheel on/off/volume knob, AM, bat **$15.00**

5203, vertical, 4x2½x1¼", 1965, blue plastic, seven transistors, upper right front round dial knob overlaps left textured grill area, left side thumbwheel on/off/volume knob, AM, bat **$15.00**

5204, vertical, 4x2½x1¼", 1965, tangerine plastic, seven transistors, upper right front round dial knob overlaps left textured grill area, left side thumbwheel on/off/volume knob, AM, bat **$15.00**

5205, vertical, 4x2½x1¼", 1965, olive plastic, seven transistors, upper right front round dial knob overlaps left textured grill area, left side thumbwheel on/off/volume knob, AM, bat $15.00

5210, vertical, 1964, eight transistors, upper right front round window dial with right side thumbwheel tuning, lower metal perforated grill area, AM, bat $20.00

5214, vertical, 1965, leatherette and chrome, 10 transistors, right front vertical slide rule dial with right side thumbwheel tuning, upper left on/off/volume window with left side thumbwheel knob, perforated grill area, strap, AM, bat **$20.00**

5217, horizontal, 1965, 10 transistors, two upper front horizontal slide rule dials, FM/AFC/AM switch, upper left thumbwheel on/off/volume knob, large lower perforated grill area, telescoping antenna, handle, AM, FM, bat **$15.00**

5219, horizontal, 1964, brown leather, eight transistors, upper left front dial, upper right on/off/volume knob, lower lattice grill area, leather handle, AM, bat **$15.00**

5220, horizontal, 1964, black leather, eight transistors, upper left front dial, upper right on/off/volume knob, lower lattice grill area, leather handle, AM, bat.................... **$15.00**

5221, horizontal, 1964, brown leather, nine transistors, upper right front dial, upper left on/off/volume knob, lower lattice grill area, leather handle, AM, bat **$15.00**

5222, horizontal, 1964, black leather, nine transistors, upper right front dial, upper left on/off/volume knob, lower lattice grill area, leather handle, AM, bat **$15.00**

5223, horizontal, 7x10¾x4", 1964, tan leather, 10 transistors, upper left front horizontal slide rule dial, three knobs, large lower grill area, leather handle, AM, bat **$15.00**

5224, horizontal, 7x10¾x4", 1964, olive leather, 10 transistors, upper left front horizontal slide rule dial, three knobs, large lower grill area, leather handle, AM, bat **$15.00**

5225, horizontal, 7x10¾x4", 1964, black leather, 10 transistors, upper left front horizontal slide rule dial, three knobs, large lower grill area, leather handle, AM, bat **$15.00**

6201, vertical, 1965, six transistors, upper right front window dial with right side thumbwheel tuning, lower grill area with horizontal bars, AM, bat **$10.00**

6214, vertical, 6⅝x3¾x2⅛", brown leatherette and chrome, 10 transistors, right front vertical slide rule dial with right side thumbwheel tuning, upper left on/off/volume window with left side thumbwheel knob, perforated grill area, strap, AM, bat $20.00

6215, vertical, 6⅝x3¾x2⅛", black leatherette and chrome, ten transistors, right front vertical slide rule dial with right side thumbwheel tuning, upper left on/off/volume window with left side thumbwheel knob, perforated grill area, strap, AM, bat **$20.00**

7228, horizontal, 1958, six transistors, lower front horizontal dial, large upper perforated grill area, handle, AM, bat **$40.00**

8220, horizontal, 4½x8x8¾", 1958, plastic, six transistors, right side dial knob, left side on/off/volume knob, front vertical grill bars, top knob rotates antenna inside case, handle, AM, bat **$50.00**

8228, horizontal/table, 1958, six transistors, step-down right side, lower right front horizontal dial, three right knobs, large perforated grill area with twin speakers, fold-down handle, AM, bat **$20.00**

9014, horizontal/table, 1959, ivory front/brown back, six transistors, raised top dial, three knobs, large lower random-patterned perforated grill area with twin speakers and lower left logo, AM, bat **$25.00**

9015, horizontal/table, 1959, ivory, six transistors, raised top dial, three knobs, large lower random-patterned perforated grill area with twin speakers and lower left logo, AM, bat **$25.00**

9016, horizontal/table, 1959, ivory front/Ming blue back, six transistors, raised top dial, three knobs, large lower random-patterned perforated grill area with twin speakers and lower left logo, AM, bat **$25.00**

9202, vertical, 1959, four transistors, upper right front half-round window dial with right side thumbwheel tuning, upper left front thumbwheel on/off/volume knob, lower lattice grill area, AM, bat **$30.00**

9204, vertical, 6¼x3¼x1½", 1959, plastic, six transistors, upper right front double window dial – upper window shows dial numbers, lower window shows CD marks – right side thumbwheel dial knob, left side thumbwheel on/off/volume knob, lower metal perforated grill area, swing handle, AM, bat **$30.00**

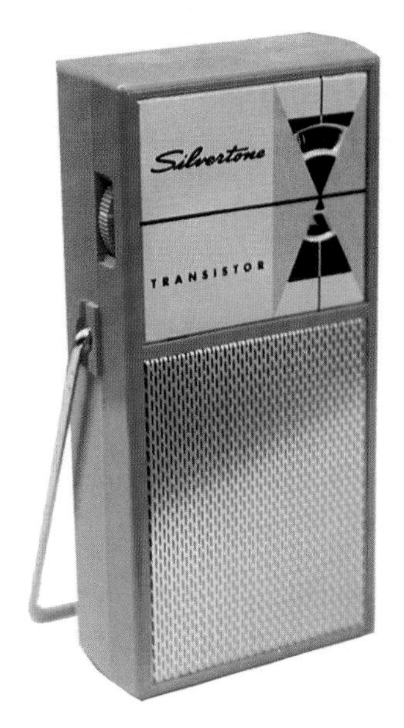

9205, (bottom left) vertical, 6¼x3¼x1½", 1959, plastic, six transistors, upper right front double window dial – upper window shows dial numbers, lower window shows CD marks – right side thumbwheel dial knob, left side thumbwheel on/off/volume knob, lower metal perforated grill area, swing handle, AM, bat ... $30.00

9222, horizontal, 6⅞x10¾x3½", 1959, tan leather, six transistors, right and left side knobs, diagonally divided front with circular grill cut-outs and upper right logo, leather handle, AM, bat **$20.00**

9226, horizontal, 1960, nine transistors, fold-back top with world map, inner multi-band slide rule dial and knobs, telescoping antenna, handle, nine bands, bat **$75.00**

Sonic

TR-500, horizontal, 1958, leather, four transistors, right and left side knobs, large front grill area with brick-shaped cut-outs, fold-down handle, AM, bat **$25.00**

Sonora

610, horizontal, 1958, five transistors, right front round dial knob, top thumbwheel on/off/volume knob, left grill area with rectangular slots, ribbed back, AM, bat **$150.00**

Sony

1R-81, horizontal, 1¾x2⅜x⅞", available in black, white, or red, eight transistors, top and bottom knobs, front metal perforated grill area with upper right logo, made in Japan, AM, bat $75.00

2F-23W, vertical, 3¼x2¼x1¼", plastic, nine transistors, solid state, upper front round two-band dial, right side thumbwheel on/off/volume knob, lower metal perforated grill area, rear FM/AM switch, fold-down telescoping antenna, made in Japan, AM, FM, bat $45.00

2R-21, square, 3x2¾x1¼", plastic/metal, eight transistors, upper right front round window dial with right side thumbwheel tuning, right side thumbwheel on/off/volume knob, large front metal circular perforated grill area, lower right side strap, made in Japan, AM, bat........ $30.00

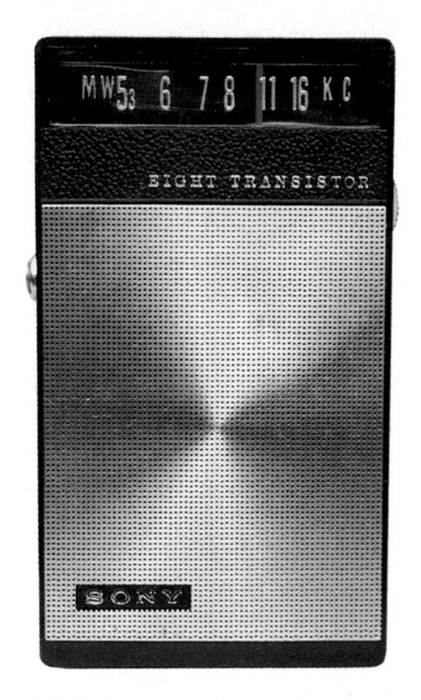

2R-27, vertical, 4x2¼x1", plastic, eight transistors, upper front horizontal slide rule dial with right side thumbwheel tuning, large lower metal perforated grill area, AM, bat $35.00

2R-30, vertical, 3⅝x2½x1¼", plastic, seven transistors, upper right front window dial with right side thumbwheel tuning, left side thumbwheel on/off/volume knob, metal perforated grill area, AM, bat **$25.00**

6R-33, horizontal, leather, nine transistors, right and left front woodgrain panels, right thumbwheel dial, left thumbwheel on/off/volume, center horizontal grill bars, H/L switch, handle, AM, bat **$15.00**

AFM-152, horizontal, 1965, 15 transistors, automatic tuning, raised top horizontal two-band slide rule dial, pushbuttons, large criss-cross grill area, telescoping antenna, handle, AM, FM, bat **$40.00**

ICR-100, horizontal, 1¼x2⅜x¾", 1966, first integrated circuit radio, top knobs, front grill area, carrying chain, plug-in recharger unit, AM, bat **$125.00**

ICR-200, horizontal, 1⅞x4⅜x1", plastic, integrated circuit radio, top left horizontal dial, top right tuning and on/off/volume knobs, front grill area with horizontal bars, lower right side strap, plug-in recharger unit, made in Japan, AM, bat $75.00

TFM-95, horizontal, 1963, available in black, cream, or turquoise, nine transistors, top horizontal two-band slide rule dial, pushbuttons, two thumbwheel knobs, large front grill area with center logo, telescoping antenna, handle, AM, FM, bat... **$30.00**

TFM-96, horizontal, 1964, nine transistors, right front round two-band dial, left perforated grill area, telescoping antenna, AM, FM, bat **$30.00**

TFM-116A, horizontal, 1964, 11 transistors, three upper front horizontal dial scales, lower perforated grill area, top pushbuttons, two telescoping antennas, handle, AM, FM, Marine, bat **$35.00**

TFM-119A, horizontal, 1965, leather, 11 transistors, upper left front horizontal three-band slide rule dial, lower grill area, top pushbuttons, telescoping antenna, leather handle, AM, FM, Marine, bat **$35.00**

TFM-121, horizontal, 1961, 12 transistors, upper front horizontal two-band slide rule dial, large lower perforated grill area with tuning knob and switch, two telescoping antennas in handle, AM, FM, bat **$35.00**

TFM-121-A, horizontal, 6⅜x9¼x2¼", 1963, plastic, 12 transistors, upper front horizontal two-band slide rule dial, large lower metal perforated grill area with tuning knob and AM/FM switch, two telescoping antennas in handle, AM, FM, bat $35.00

TFM-151, vertical, 1960, 15 transistors, top raised horizontal dial, top right and left knobs, front perforated grill area, telescoping antenna, handle, AM, FM, bat **$90.00**

TFM-825, vertical, 1964, eight transistors, left front vertical two-band slide rule dial, step-down top with two thumbwheel knobs, telescoping antenna, AM, FM, bat **$25.00**

TFM-917W, horizontal, 1965, leather, nine transistors, upper front horizontal two-band slide rule dial, large lower grill area with horizontal bars, telescoping antenna, handle, AM, FM, bat **$25.00**

TFM-951, horizontal, 1964, nine transistors, top horizontal two-band slide rule dial, pushbuttons, two thumbwheel knobs, large front grill area with center logo, telescoping antenna, handle, AM, FM, bat **$30.00**

TR-55, horizontal, 1955, Sony's first transistor radio, upper right front dial over large perforated grill area, lower right side thumbwheel knob, AM, bat .... **$600.00+**

TR-63, (bottom left) vertical, 4¼x2¾ x1¼", 1957, available in red, black, green, or yellow plastic, six transistors, Sony's first imported pocket radio, upper left front round dial knob, right side thumbwheel on/off/volume knob, lower metal perforated grill area, AM, bat $300.00

TR-84 "Super Sensitivity," horizontal, 4x7¼x1¾", 1961, available in gray, beige, cream, or green plastic, eight transistors, upper right front dial with thumbwheel tuning, lower right front on/off/volume window with thumbwheel knob, left grill area with horizontal bars, made in Japan, AM, bat $35.00

TR-86, vertical, 4½x2⅞x1⅜", 1959, plastic, eight transistors, upper right front round dial with right side thumbwheel tuning, left on/off/volume window with thumbwheel knob, lower metal perforated grill area, made in Japan, AM, bat **$175.00**

TR-510, vertical, 1961, five transistors, upper front window dial with thumbwheel tuning, lower round perforated grill area, swing handle, AM, bat **$100.00**

TR-608, horizontal, 1961, six transistors, right half-round dial with thumbwheel tuning, lower right side thumbwheel on/off/volume knob, left grill area with horizontal bars, AM, bat $50.00

TR-609, horizontal, 3¾x6x1½", 1962, plastic, six transistors, right front round dial over large lattice grill area, top left thumbwheel on/off/ volume knob, AM, bat $60.00

TR-610, vertical, 4¼x2¾x1¼", 1959, available in black, red, green, or ivory plastic, six transistors, upper front window dial with right side thumbwheel tuning, right side thumbwheel on/off/volume knob, lower round metal perforated grill area, swing handle, AM, bat $100.00

TR-620, vertical, 3½x2⅜x1", 1961, plastic, six transistors, upper left front window dial with right side thumbwheel tuning, right side thumbwheel on/off/volume knob, round metal convex perforated grill area, left side strap, made in Japan, AM, bat $70.00

TR-624, horizontal/desk-top radio, 1962, available in brown or black, six transistors, flip-up front, inner thumbwheel knobs and grill, radio plays when lid is opened and shuts off when lid is closed, AM, bat **$50.00**

TR-627, horizontal, 8⅛x11x3⅜", plastic, six transistors, upper front horizontal slide rule dial, three knobs, horizontal grill bars, handle, feet, made in Japan, AM, bat $30.00

TR-630, vertical, 3½x2½x1", 1963, plastic, six transistors, upper front window dial with right side thumbwheel tuning, right side thumbwheel on/off/volume knob, perforated grill area, AM, bat **$35.00**

TR-650, vertical, 1963, six transistors, upper left front window dial with thumbwheel tuning, large round grill area, AM, bat **$40.00**

TR-712, horizontal, 1961, seven transistors, right front dial, lower right on/off/volume window with right side knob, left checkered grill area, handle, AM, bat **$40.00**

TR-714, horizontal, 3x4½x1⅜", 1961, seven transistors, rounded top right with two thumbwheel knobs, upper front horizontal two-band dial, lower metal perforated grill area with switch, telescoping antenna, AM, SW, bat $65.00

TR-717Y, horizontal, 1962, seven transistors, center front round two-band dial, two right knobs, left lattice grill area, AM, SW, bat **$35.00**

TR-725, horizontal, 1963, available in gray or ivory, seven transistors, upper right front window dial with thumbwheel tuning, large left round perforated grill area, telescoping antenna, AM, SW, bat **$45.00**

TR-730, vertical, 1963, available in black or bone white, seven transistors, upper left front window dial with thumbwheel tuning, large lower perforated grill area, AM, bat...... **$30.00**

TR-733, horizontal, 1964, seven transistors, upper front horizontal two-band slide rule dial, two right side thumbwheel knobs, round grill area, telescoping antenna, AM, SW, bat **$30.00**

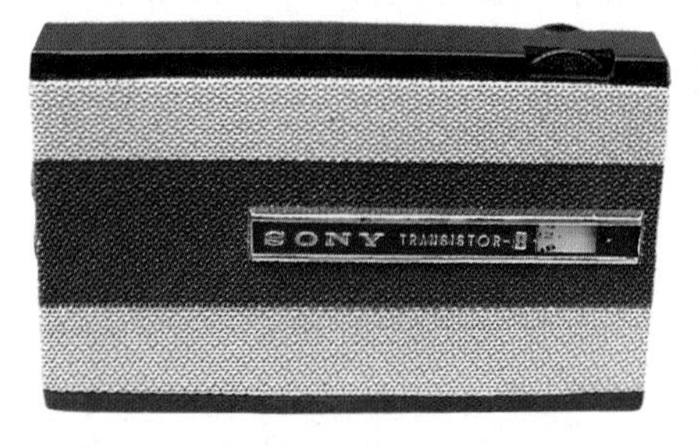

TR-810, horizontal, 3x5¼x1", 1961, plastic, eight transistors, large front metal perforated grill area with right window dial, two top right thumbwheel knobs, AM, bat $45.00

TR-812, horizontal, 1961, eight transistors, upper front horizontal three-band slide rule dial, large lower perforated grill area, telescoping antenna, fold-down handle, AM, 2SW, bat **$30.00**

TR-814, horizontal, 1961, eight transistors, large right front round three-band dial, recessed right with upper thumbwheel on/off/volume knob and lower band switch, telescoping antenna, handle, AM, 2SW, bat **$35.00**

TR-817, vertical, 1963, eight transistors, three upper front round windows – right on/off/volume, center tuning meter, left dial – top on/off button, right side thumbwheel knobs, lower perforated grill area, AM, bat $50.00

TR-818, horizontal, 1963, available in gray or cream, eight transistors, upper front horizontal slide rule dial with thumbwheel tuning, lower left grill area with rectangular cutouts, AM, bat **$30.00**

TR-826, vertical, 1964, eight transistors, upper front horizontal slide rule dial with thumbwheel tuning, large lower perforated grill area, AM, bat...................... **$25.00**

TR-881, horizontal, 4¾x7¾x2", plastic, eight transistors, two top horizontal dials, right and left side knobs, large front metal perforated grill area, telescoping antenna, AM, Marine, bat **$40.00**

TR-1811, vertical, 1965, six transistors, upper left front round dial with right side thumbwheel tuning, right volume window with right side thumbwheel knob, lower perforated grill area, AM, bat **$25.00**

TR-1820, horizontal, 3¼x5⅞x1⅞", plastic, six transistors, right front round dial knob, top left thumbwheel on/off/volume knob, left horizontal grill bars, made in Japan, AM, bat $20.00

TR-6080, 4⅜x7⅞x1⅞", horizontal, 1963, available in green, red, or ivory plastic, six transistors, upper right front window dial with right side thumbwheel tuning, lower right side thumbwheel on/off/volume knob, left round grill area with horizontal bars, swing handle, made in Japan, AM, bat $30.00

TR-6120, horizontal, 1964, six transistors, large right front dial, lower right side on/off/volume knob, left grill area with horizontal slots, base, handle, AM, bat **$20.00**

TR-7120, horizontal, 6¾x11¼x4⅜", 1962, plastic, seven transistors, large right front dial, lower right side on/off/volume knob, left checkered grill area, base, handle, AM, bat ... **$25.00**

TRW-621, vertical/watch radio, 4⅛x2⅝x1⅛", 1962, available in black, gray, or beige plastic, six transistors, right window dial with right side thumbwheel tuning, upper right front thumbwheel on/off/volume knob, left watch face with top left winding stem, top right "auto/manu" switch, lower metal perforated grill area with center logo, swing handle, made in Japan, AM, bat $100.00

Sorrento

T-666, vertical, 4¼x2⅝x1¼", plastic, six transistors, two upper front windows – right dial, left on/off/volume – right and left side thumbwheel knobs, lower metal perforated grill area, made in Japan, AM, bat $25.00

SounDesign

SD-1094, vertical, 1965, 10 transistors, upper right front window dial with thumbwheel tuning, large lower lattice grill area, AM, bat **$10.00**

SD-1670, vertical, 1965, six transistors, upper right front window dial with thumbwheel tuning, lower grill area with horizontal bars, AM, bat **$15.00**

SD-2091, horizontal, 1965, 10 transistors, upper front horizontal slide rule dial, large lower perforated grill area with lower right FM/AM switch, telescoping antenna, AM, FM, bat **$15.00**

Spica

ST-600, horizontal, 1965, plastic, six transistors, right front round dial knob, top thumbwheel on/off/volume knob, left horseshoe-shaped grill area with rectangular cut-outs, AM, bat $35.00

Sportmaster

47900, vertical, 1965, six transistors, upper right front window dial with right side thumbwheel tuning, large lower grill area with horizontal bars, AM, bat **$15.00**

47915, vertical, 1965, eight transistors, upper right front window dial with thumbwheel tuning, large lower grill area with vertical bars, AM, bat **$10.00**

47920, vertical, 1965, nine transistors, upper right front window dial with thumbwheel tuning, large lower grill area with horizontal bars, AM, bat **$10.00**

47925, vertical, 1965, 10 transistors, upper right front window dial with thumbwheel tuning, large lower lattice grill area, AM, bat **$10.00**

47965, horizontal, 1965, 10 transistors, upper front horizontal slide rule dial, large lower perforated grill area with lower right FM/AM switch, telescoping antenna, AM, FM, bat **$15.00**

Standard

SR-F22, vertical, 1959, six transistors, upper front round dial knob, upper right thumbwheel on/off/volume knob, lower perforated grill area with vertical divider and lower right logo, AM, bat **$45.00**

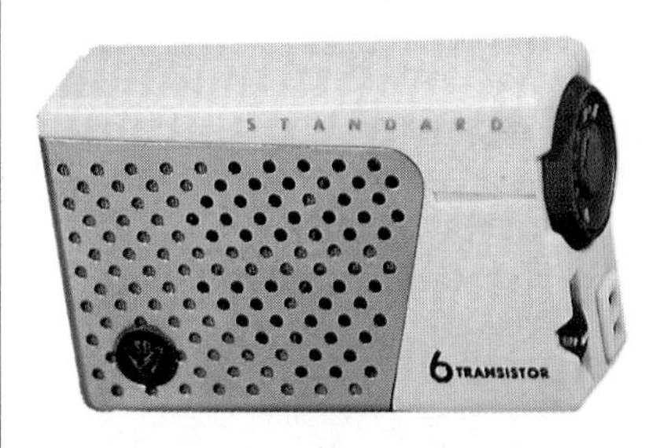

SR-F25, horizontal, plastic, six transistors, upper right side dial knob, lower right side thumbwheel on/off/volume knob, large left front metal grill area with circular cut-outs and lower left logo, AM, bat $30.00

SR-F415, vertical, 1965, six transistors, upper right front window dial with right side thumbwheel tuning, large lower grill area with vertical slots, AM, bat **$15.00**

SR-G45 "Micronic Ruby," square, 1⅞x1⅞x⅞", 1966, plastic, two right side knobs – upper on/off/volume, lower tuning – front metal perforated grill area, left side strap, made in Japan, AM, bat $75.00

SR-G430 "Micronic Ruby," square, 1⅞x1⅝x¾", 1964, plastic, seven transistors, right side tuning and on/off/volume knobs, front metal perforated grill area with center logo, left side strap, AM, bat $85.00

SR-G433 "Micronic Ruby," square, 1965, seven transistors, right side tuning and on/off/volume knobs, front grill area with horizontal slots, left side vinyl strap, AM, bat........ $75.00

SR-H436 "Micronic Ruby," horizontal, 1⅝x2¼x1", plastic, eight transistors, two right side knobs – upper on/off/volume, lower tuning – front metal perforated grill area, left side strap, made in Japan, AM, bat $65.00

SR-H437 "Micronic Ruby," square, 2x1⅞x1", 1964, plastic, eight transistors, two right side knobs – upper on/off/volume, lower tuning – lower front metal perforated grill area, left side metal chain and medallion, AM, bat **$75.00**

SR-H438, horizontal, 1965, eight transistors, right side tuning and on/off/volume knobs, front grill area with horizontal slots, AM, bat $65.00

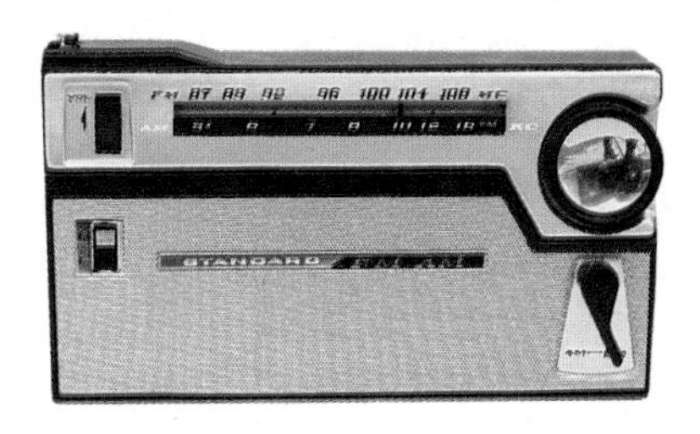

SR-J100F, horizontal, 5x8¾x2¼", 1962, plastic, 10 transistors, upper front horizontal two-band slide rule dial with right dial knob, upper left thumbwheel on/off/volume knob, large lower metal perforated grill area with left H/L switch and right AM/FM switch, telescoping antenna, made in Japan, AM, FM, bat... $30.00

SR-J715F, horizontal, 1964, 10 transistors, upper front horizontal FM slide rule dial, right side thumbwheel knob, lower right AFC switch, telescoping antenna, FM, bat...... **$25.00**

SR-J716F, horizontal, 1964, 10 transistors, upper front horizontal two-band slide rule dial, right side thumbwheel knob, lower right AM/FM switch, telescoping antenna, AM, FM, bat **$25.00**

Star-Lite

AP-642, horizontal, 1964, nine transistors, upper right front round two-band dial with right side thumbwheel tuning, right side thumbwheel on/off/volume knob, perforated grill area, telescoping antenna, strap, AM, FM, bat **$25.00**

"Boy's Radio," vertical, 4x2½x1⅛", plastic, two transistors, upper front window dial with left side thumbwheel tuning, right side thumbwheel on/off/volume knob, round metal perforated grill area, made in Japan, AM, bat **$30.00**

DE-62 "HiFi Deluxe," vertical, 3½ x2¼x1", plastic, six transistors, upper right front window dial with right side thumbwheel tuning, left side thumbwheel on/off/volume knob, metal perforated grill area, AM, bat $20.00

DP-118, vertical, 1965, six transistors, upper front window dial with right side thumbwheel tuning, lower lattice grill area, AM, bat **$15.00**

DP-222 "Leatherneck," horizontal, 1965, leather, 12 transistors, right front dial knob, left on/off/volume knob, center grill area with vertical bars, leather handle, AM, bat **$15.00**

FM-500 "Discoverer," horizontal, 1965, 12 transistors, upper front horizontal five-band slide rule dial with right thumbwheel tuning, large lower lattice grill area with lower right logo, telescoping antenna, handle, AM, FM, 3SW, bat **$35.00**

FM-900, horizontal, 1965, nine transistors, upper front horizontal two-band slide rule dial, large lower perforated grill area with logo, telescoping antenna, AM, FM, bat **$15.00**

PM-714, horizontal, 1965, leather, 10 transistors, upper front horizontal two-band slide rule dial, three right knobs, large perforated grill area with lower left logo, telescoping antenna, AM, FM, bat **$15.00**

T-603, vertical, plastic, six transistors, upper front window dial with right side thumbwheel tuning, upper left front thumbwheel on/off/volume knob, lower metal perforated grill area, AM, bat **$35.00**

TD-660 "Duke," vertical, 1965, 10 transistors, upper front window dial with thumbwheel tuning, lower crisscross grill area, AM, bat **$15.00**

TM-680, vertical, 1965, 10 transistors, upper front window dial with thumbwheel tuning, lower six-section round grill area, AM, bat **$15.00**

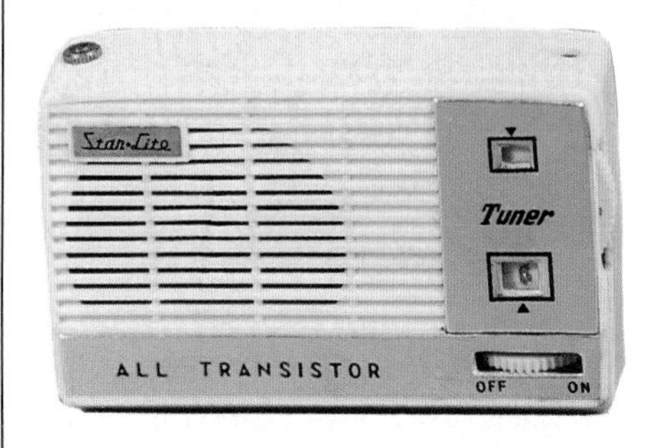

TR-21, horizontal, plastic, two right front dial windows with right side thumbwheel tuning, lower right front thumbwheel on/off/volume knob, left lattice grill area with upper left logo, AM, bat $30.00

TR-709 "Voyager," vertical, 1965, six transistors, upper right front window dial with right side thumbwheel tuning, lower grill area with concentric square pattern, AM, bat **$15.00**

TR-960 "Rough Rider," horizontal, 1965, leather, 12 transistors, upper right front round dial, lower on/off/volume knob, left lattice grill area, handle, AM, bat **$15.00**

TRJ-10 "High Fidelity," vertical, 1965, 10 transistors, upper front window dial with thumbwheel tuning and wedge-shaped trim, lower grill area with horizontal slots, AM, bat **$25.00**

TRN-112 "High Sensitivity," horizontal, 1964, 12 transistors, upper right front half-round dial, large lower perforated grill area, handle, AM, bat $20.00

TS-640 "Skymate," vertical, 1965, 10 transistors, upper front oval window dial with thumbwheel tuning, lower grill area with criss-cross pattern, AM, bat.................. $20.00

Sudfunk

K986A, horizontal, 1961, nine transistors, rounded case, right front round two-band dial, three top pushbuttons, left grill area with vertical bars, telescoping antenna, handle, AM, FM, bat............. $35.00

Summit

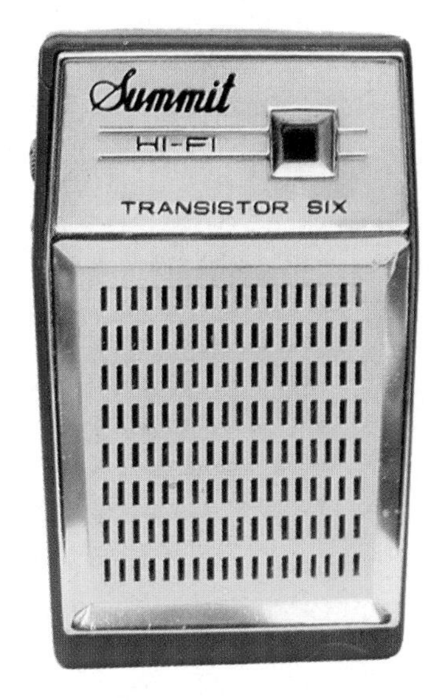

HS-657 "HiFi," vertical, 4⅛x2½x1⅛", plastic, six transistors, upper right front window dial with right side thumbwheel tuning, left side thumbwheel on/off/volume knob, metal grill area with vertical slots, made in Ryukyu, AM, bat $20.00

S109, vertical, 4½x2½x1⅜", plastic, 10 transistors, upper right front window dial with right side thumbwheel tuning, left side thumbwheel on/off/volume knob, oval metal perforated grill area, made in Ryukyu, AM, bat $30.00

Suntone

1112, horizontal, 2⅝x4½x1¼", plastic, solid state, upper right front window dial with right side thumbwheel tuning, lower right front on/off/volume window with right side thumbwheel knob, left textured grill area, made in Hong Kong, AM, bat $20.00

Super

TR-261, vertical, plastic, two transistors, upper right front window dial with right side thumbwheel tuning, left side thumbwheel on/off/volume knob, metal perforated grill area, AM, bat $30.00

Superex

TR-66, vertical, 1960, six transistors, upper front round dial knob, lower perforated grill area with small on/off/volume knob, swing handle, AM, bat **$30.00**

Supertone

AR800, vertical, 5x3⅜x1", plastic, book-style radio, upper front window dial with "king and queen" pictures, lower grill area with horizontal slots, two thumbwheel knobs – one tuning, one on/off/volume – on left "spine" of book, made in Japan, AM, bat $35.00

Supreme

TR-803, vertical, 3⅜x2½x1", plastic with metal side trim, seven transistors, upper front window dial with right front thumbwheel tuning, left front thumbwheel on/off/volume knob, round metal perforated grill area, AM, bat **$55.00**

TR-861, vertical, 3⅞x2⅝x1¼", 1962, plastic, six transistors, upper right front round window dial with right side thumbwheel tuning, left side thumbwheel on/off/volume knob, round metal perforated grill area, made in Japan, AM, bat......... **$45.00**

Sutton

J683, vertical, plastic, seven transistors, upper right front window dial

with right side thumbwheel tuning, top left thumbwheel on/off/volume knob, lower oval grill area with vertical bars, AM, bat $10.00

Sylvania

4P06E, vertical, 1962, four transistors, upper right front window dial with right side thumbwheel tuning, left side thumbwheel on/off/volume knob, lower perforated grill area with lower left logo, AM, bat **$30.00**

4P14, horizontal, 1961, four transistors, upper left front window dial with thumbwheel tuning, large lower grill area with lower right logo and lower left knob, AM, bat **$50.00**

4P19W, horizontal, 3½x6½x1¾", 1962, plastic, four transistors, upper right front window dial with thumbwheel tuning, lower right side thumbwheel on/off/volume knob, large left grill area with circular cut-outs, AM, bat **$30.00**

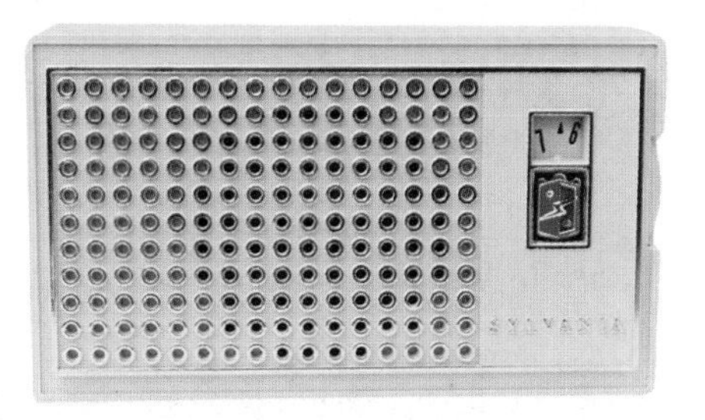

4P19WD, horizontal, 3½x6½x1¾", 1962, plastic, four transistors, upper right front window dial with thumbwheel tuning, lower right side thumbwheel on/off/volume knob, large left grill area with circular cut-outs, AM, bat $30.00

5P11R, horizontal, 1960, five transistors, upper left front window dial with top thumbwheel tuning, lower left on/off/volume knob, horizontal grill bars with right logo, AM, bat **$35.00**

6P09T, vertical, 1961, six transistors, upper right front window dial with right side thumbwheel tuning, left side thumbwheel on/off/volume knob, lower perforated grill area with logo, AM, bat **$25.00**

7K10, horizontal/clock radio, 1961, seven transistors, lower right front dial knob overlaps lattice grill area, left front alarm clock face and on/off/volume knob, swing handle, AM, bat **$25.00**

7P12T, vertical, 1960, seven transistors, upper right front window dial, upper left on/off/volume knob, large lower perforated grill area with lower left logo and right vertical bar, swing handle, AM, bat **$30.00**

7P13, horizontal, 1960, leather, seven transistors, right front dial knob over large perforated grill area with upper left logo, top left knob, leather handle, AM, bat..................... **$20.00**

2700 "Golden Shield," horizontal, 3½ x6½x1¾", leather, five transistors, right front round dial knob, large metal perforated grill area, leather handle, AM, bat $20.00

2808 "Golden Shield," (bottom left) vertical, 6¾x4¼x2", 1960, plastic, upper right round dial with right side thumbwheel tuning, upper left gold shield logo, left side thumbwheel on/off/volume knob, lower metal perforated grill area, pull-up handle, AM, bat $25.00

3102 "Thunderbird," 1957, resembles Ford Thunderbird car, two front thumbwheel knobs, lift top, inner clear plastic cover over radio chassis, top built-in speaker, handle, AM, bat **$225.00**

3204TU, horizontal, 1958, six transistors, right front round dial, top left on/off/volume knob, large lattice grill area with lower right "T6" logo, fold-down handle, AM, bat **$35.00**

3305BL, horizontal, 6¾x9¼x3¾", 1958, navy/white plastic, six transistors, right front round dial, top left on/off/volume knob, large lattice grill area with lower right "T6" logo, fold-down handle, AM, bat **$35.00**

3305TA, horizontal, 6¾x9¼x3¾", 1958, red/white plastic, six transistors, right front round dial, top left on/off/volume knob, large lattice grill area with lower right "T6" logo, fold-down handle, AM, bat **$35.00**

3406 Series, horizontal/clock radio, 1960, leather, seven transistors, lower front thumbwheel dial and on/off/volume knobs overlap woven grill area with upper left logo, right alarm clock face, AM, bat **$25.00**

TH16 Series, vertical, 1963, eight transistors, upper right front dial with right side thumbwheel tuning, upper left front on/off/volume knob, lower right perforated grill area, lower left logo, swing handle, AM, bat **$25.00**

TR-22, horizontal, 1964, seven transistors, right front window dial with right side thumbwheel tuning, left perforated grill area with lower left logo, AM, bat **$15.00**

TR-25, vertical, 1964, eight transistors, upper front horizontal slide rule dial with thumbwheel tuning, lower perforated grill area with lower left logo, AM, bat **$15.00**

TR35, horizontal, 1964, nine transistors, right front vertical two-band dial, three knobs, left perforated grill area with lower left logo, two top pushbuttons, telescoping antenna, AM, FM, bat **$15.00**

TR50, vertical, 1965, six transistors, upper front round dial with thumbwheel tuning, lower perforated grill area with lower left logo, AM, bat **$20.00**

TR54, vertical, 4x2½x1¼", 1965, eight transistors, upper front horizontal slide rule dial with right side thumbwheel tuning, lower perforated grill area with lower left logo, AM, bat **$15.00**

TR58, horizontal, 1965, eight transistors, vertical slide rule dial, two right front knobs – upper tuning, lower on/off/volume – left grill area, handle, AM, bat **$15.00**

TR62, horizontal, 1965, nine transistors, two right front window dials – upper FM, lower AM – left perforated grill area, telescoping antenna, fold-down handle, AM, FM, bat **$15.00**

Symphonic

S-62, vertical, 1963, six transistors, upper right front round window dial over large perforated grill area, right side thumbwheel tuning knob, top left thumbwheel on/off/volume knob, AM, bat **$15.00**

S-73, horizontal, 1963, seven transistors, right front window dial with right side thumbwheel tuning, left perforated grill area with lower left logo, AM, bat **$20.00**

S-84, vertical, 1963, eight transistors, upper front half-round window dial with thumbwheel tuning, lower perforated grill area with logo, AM, bat **$15.00**

S-93, vertical, 1963, nine transistors, upper front window dial with thumbwheel tuning over large perforated grill area with lower left logo, AM, bat **$15.00**

SF-400, horizontal, 1963, nine transistors, right front three-band dial knob over large checkered grill area with lower left logo, upper left thumbwheel on/off/volume knob, SW/AM/FM switch, AM, FM, SW, bat **$20.00**

Tact

TPR-61 "Phonoradio," vertical, 7x3¾x1¾", six transistors, combination phonograph and radio, plays 45s and 33s with built-in tone arm, AM, bat **$75.00**

Tama-Tone

"Boy's Radio," vertical, plastic, two transistors, upper right front window dial with right side thumbwheel tuning, left side thumbwheel on/off/volume knob, lower lattice grill area, AM, bat $30.00

Telefunken

"Ticcolo," horizontal/watch radio, 3 x5¼x1½", six transistors, upper front horizontal two-band slide rule dial with right and left thumbwheel knobs, lower right watch face, left metal perforated grill area, MW, LW, bat **$50.00**

Tempest

HT-1251, vertical, 4¼x2⅝x1¼", plastic, 14 transistors, upper right front window dial with right side thumbwheel tuning, top left thumbwheel on/off/volume knob, two oval metal perforated grill areas, made in Japan, AM, bat $30.00

HT-8041 "Deluxe," vertical, 4¼x2⅝x 1¼", plastic, eight transistors, upper front oval window dial with right side thumbwheel tuning, top left thumbwheel on/off/volume knob, lower metal perforated grill area, made in Japan, AM, bat **$25.00**

TR 1200, vertical, 4¼x2⅝x1¼", plastic, upper right round dial window with right side thumbwheel tuning, upper left round on/off/volume window with left side thumbwheel knob, lower metal perforated grill area with lower right

logo, made in Hong Kong, AM, bat $15.00

Terra

5026, (bottom left) vertical, 4⅜x2¾ x 1¼", plastic, 12 transistors, upper right front window dial with right side thumbwheel tuning, left side thumbwheel on/off/volume knob, strap, made in Japan, AM, bat. $10.00

Tokai

FA-951, horizontal, 1963, nine transistors, off-center vertical two-band slide rule dial, two top thumbwheel knobs, two lower switches, left perforated grill area, telescoping antenna, AM, FM, bat **$20.00**

HA-911 "Super Sensitivity," horizontal, 1964, nine transistors, upper front horizontal dial, three thumbwheel knobs, large perforated grill area with lower left logo, AM, bat **$15.00**

RA-9, horizontal, 1965, nine transistors, upper right front window dial with right side thumbwheel tuning, lower right side thumbwheel on/off/volume knob, horizontal grill bars, AM, bat **$15.00**

RA-711, vertical, 1963, seven transistors, upper right front dial with thumbwheel tuning, left side thumbwheel on/off/volume knob, large perforated grill area, AM, bat **$25.00**

RA-801, horizontal, 1964, eight transistors, right front window dial with thumbwheel tuning, lower right side thumbwheel on/off/volume knob, left perforated grill area with upper left logo, upper right "L/H" switch, AM, bat **$25.00**

Tonecrest

645, vertical, 1965, six transistors, upper left front window dial with thumbwheel tuning, right side thumbwheel on/off/volume knob, large grill area with horizontal bars, AM, bat **$10.00**

946, vertical, 1965, nine transistors, upper right front window dial with thumbwheel tuning, large lower grill area with vertical bars, AM, bat **$10.00**

1051, horizontal, 1965, 10 transistors, right front vertical two-band slide rule dial with thumbwheel tuning, left grill area with horizontal bars, telescoping antenna, AM, FM, bat **$15.00**

1094, vertical, 1965, 10 transistors, upper right front window dial with thumbwheel tuning, large lower lattice grill area, AM, bat **$10.00**

1670, vertical, 1965, six transistors, upper right front window dial with thumbwheel tuning, lower grill area with horizontal bars, AM, bat **$15.00**

1889, vertical, 1965, eight transistors, upper right front window dial with thumbwheel tuning, left side thumbwheel on/off/volume knob, lower grill area with vertical bars, AM, bat **$10.00**

2091, horizontal, 1965, 10 transistors, upper front horizontal slide rule dial, large lower perforated grill area with lower right FM/AM switch, telescoping antenna, AM, FM, bat **$15.00**

Top-Flight

"Boy's Radio," vertical, 4⅛x2⅝x1¼", plastic, two transistors, upper right front round window dial forms open mouth of lion's head decoration, right side thumbwheel tuning knob, left on/off/volume window with left side thumbwheel knob, lower metal perforated grill area, made in Japan, AM, bat $40.00

Toshiba

3TP-315Y, vertical, 1959, three transistors, upper right front thumbwheel dial, upper left thumbwheel on/off/volume knob, lower horizontal grill bars, no built-in speaker, earphone only, AM, bat.......................... **$75.00**

5TP-90, vertical, 4x2½x1", 1961, plastic, five transistors, upper right front window dial with right side thumbwheel tuning, left side thumbwheel on/off/volume knob, metal perforated grill area with lower left logo, AM, bat $30.00

6P-10, horizontal, 1963, six transistors, right front window dial with right side thumbwheel tuning, right side thumbwheel on/off/volume knob, left perforated grill area with center logo, AM, bat **$25.00**

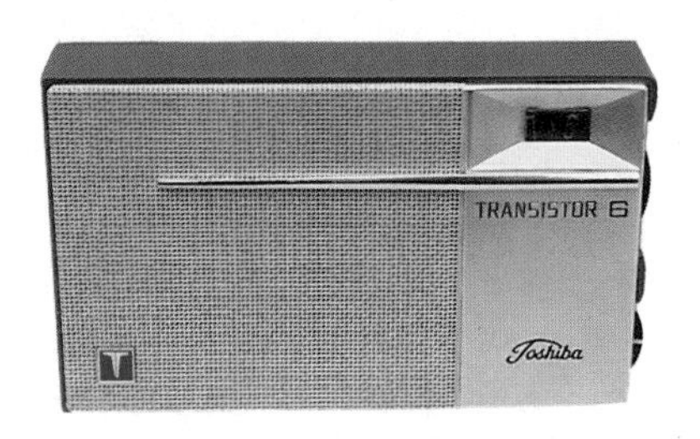

6P-15, horizontal, 2⅝x4½x1⅛", 1962, plastic, six transistors, upper right front window dial with right side thumbwheel tuning, lower right side thumbwheel on/off/volume knob, left metal perforated grill area with lower left logo, AM, bat $30.00

6P-35, horizontal, 2⅝x4⅛x1⅛", plastic, six transistors, upper right front window dial with right side thumbwheel tuning, lower right side thumbwheel on/off/volume knob, left metal perforated grill area, left side strap, made in Japan, AM, bat $25.00

6TC-485, vertical/folding-style clock radio, 4¾x2⅞x2¼" (closed), 1963, leatherette, six transistors, inner right round dial/thumbwheel knobs/metal perforated grill area, inner left round alarm clock face and lower knob, AM, bat **$50.00**

6TC-485A, vertical/folding-style clock radio, 4¾x2⅞x2¼" (closed), 1963, leatherette, six transistors, inner right round dial/thumbwheel knobs/metal perforated grill area, inner left round alarm clock face and lower knob, AM, bat **$50.00**

6TP-31, vertical, 4½x2¾x1¼", plastic, six transistors, upper front half-round dial panel overlaps large metal perforated grill area, right side thumbwheel tuning, lower left logo, rear fold-out stand, AM, bat **$75.00**

6TP-31A, vertical, 4½x2¾x1¼", 1963, plastic, six transistors, upper front half-round dial panel overlaps large metal perforated grill area, right side thumbwheel tuning, lower left logo, rear fold-out stand, AM, bat **$75.00**

6TP-304, vertical, 1959, six transistors, upper left front vertical slide rule dial with thumbwheel tuning, lower right perforated grill area with lower right logo, AM, bat **$45.00**

6TP-309, vertical, 1959, six transistors, right front window dial inside V-shaped trim, right side thumbwheel tuning, left side thumbwheel on/off/volume knob, lower perforated grill area, AM, bat **$75.00**

6TP-309Y, vertical, 1959, six transistors, right front window dial inside V-shaped trim, right side thumbwheel tuning, left side thumbwheel on/off/volume knob, lower perforated grill area, AM, bat **$75.00**

6TP-314, vertical, 4½x2¾x1¼", 1959, available in white, green, or coral plastic, six transistors, upper left front window dial with right side thumbwheel tuning, right side thumbwheel on/off/volume knob, horizontal grill bars with lower right logo, AM, bat **$30.00**

6TP-354, vertical, 1960, six transistors, upper right front thumbwheel dial, upper left thumbwheel on/off/volume knob, lower perforated grill area, AM, bat.......................... **$45.00**

6TP-357, vertical, 1961, six transistors, upper right front round dial knob, left side thumbwheel on/off/volume knob, lower perforated grill area, AM, bat.......................... **$60.00**

6TP-385, horizontal, 2½x4¼x1¼", 1961, plastic, six transistors, right front control panel with window dial and right thumbwheel tuning, metal perforated grill area, telescoping antenna, AM, bat $40.00

6TP-394, vertical, 2⅞x2⅜x1⅛", 1961, plastic, six transistors, upper right front round dial knob, upper left side thumbwheel on/off/volume knob, lower metal perforated grill area, AM, bat $45.00

6TR-92, round, 1959, six transistors, spherical-shaped set with floral design, top dial, bottom speaker grill, base, handle, AM, bat **$200.00**

6TR-186, horizontal, 1959, six transistors, right front thumbwheel dial and thumbwheel on/off/volume knob, left lace grill area, AM, bat **$100.00**

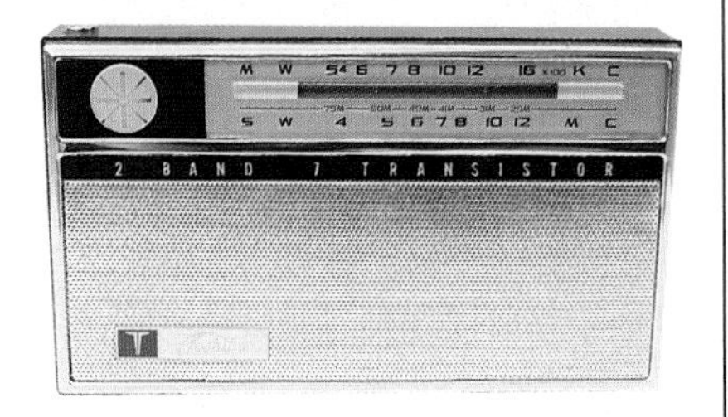

7P-130S, horizontal, 3¼x5⅞x1½", 1963, plastic, seven transistors, upper front horizontal two-band slide rule dial with right side thumbwheel tuning, right side thumbwheel on/off/volume knob, lower right side thumbwheel fine tuning knob, metal perforated grill area, rear SW/MW switch, telescoping antenna, made in Japan, AM, SW, bat $35.00

7TH-425, round, 12x4", 1961, plastic, seven transistors, made to hang on the wall, center round two part dial – the outside is for tuning and the inside is for volume control – on/off switch is a pull-cord at bottom of case, two speakers, AM, bat $200.00

7TH-425Y, round, 12x4", 1961, plastic, seven transistors, made to hang on the wall, center round two part dial – the outside is for tuning and the inside is for volume control – on/off switch is a pull-cord at bottom of case, two speakers, AM, bat **$200.00**

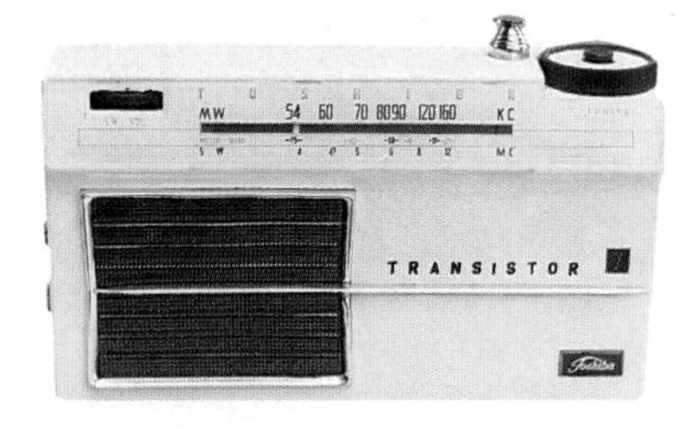

7TM-312S, horizontal, 4x7x1¾", 1961, seven transistors, plastic, upper front horizontal two-band slide rule dial with top right tuning knob, upper left front thumbwheel knob, lower left metal perforated grill area with center horizontal bar, telescoping antenna, AM, SW, bat $50.00

7TP-21, vertical, 1962, seven transistors, upper right front round window dial with right side thumbwheel tuning, upper left thumbwheel on/off/volume knob, center perforated grill area, AM, bat **$40.00**

7TP-30, vertical, 1962, seven transistors, upper right front round window dial with right side thumbwheel tuning, upper left thumbwheel on/off/volume knob, center perforated grill area, AM, bat **$40.00**

7TP-303, vertical, 4⅜x2¾x1⅜", 1961, plastic, seven transistors, upper front "cat's eye" window dial with right side thumbwheel tuning, right side thumbwheel on/off/volume knob, lower metal perforated grill area with V-shaped decoration, could be used with optional speaker box model 3WX, made in Japan, AM, bat.
radio without speaker box $75.00
radio with speaker box $125.00

7TP-352M, vertical, 5x3x1¼", 1961, plastic, seven transistors, upper front horizontal two-band slide rule dial with right thumbwheel tuning, right side thumbwheel on/off/volume knob, lower metal perforated grill area, top left band switch, telescoping antenna, made in Japan, AM, Marine, bat **$40.00**

7TP-352S, vertical, 5x3x1¼", 1961, plastic, seven transistors, upper front horizontal two-band slide rule dial with right thumbwheel tuning, right side thumbwheel on/off/volume knob, lower metal perforated grill area, top left band switch, telescoping antenna, made in Japan, AM, SW, bat $40.00

8TH-428R, horizontal/table, 1963, eight transistors, three right front horizontal slide rule dial scales, three knobs, left checkered grill area, feet, AM, 2SW, bat **$25.00**

8TM-41, horizontal, 3¾x6½x1½", 1962, plastic, eight transistors, upper

front slanted horizontal dial, right side tuning knob, left side on/off/ volume knob, metal perforated grill area, AM, bat $40.00

8TM-210S, horizontal, plastic, eight transistors, upper right front two-band window dial with thumbwheel tuning, upper left thumbwheel on/ off/volume knob, large perforated grill area, AM, SW, bat $50.00

8TM-294, horizontal, 3¾x6½x1½", 1960, plastic, eight transistors, top horizontal wrap-over dial with right side tuning knob, left side on/off/ volume knob, lower metal grill area with rectangular cut-outs, made in Japan, AM, bat **$50.00**

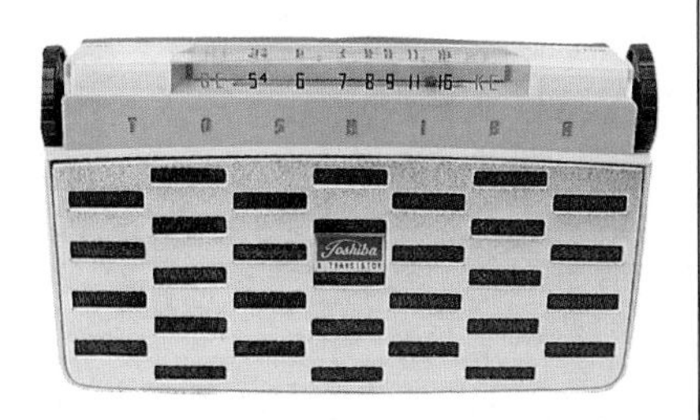

8TM-294B, horizontal, 3¾x6½x1½", 1960, plastic, eight transistors, top horizontal wrap-over dial with right side tuning knob, left side on/off/ volume knob, lower metal grill area with rectangular cut-outs, made in Japan, AM, bat $50.00

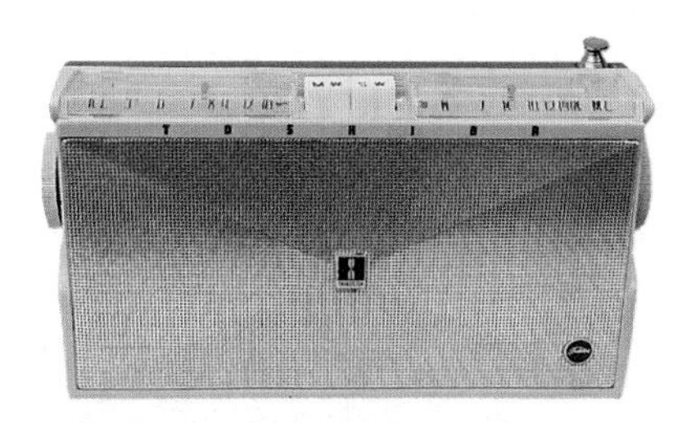

8TM-300S, horizontal, 4⅝x8x1¾", 1960, plastic, eight transistors, two top horizontal dials, right side tuning knob, left side on/off/volume knob, top MW/SW pushbuttons, large metal perforated grill area with center logo, telescoping antenna, made in Japan, AM, SW, bat $35.00

8TM-613, horizontal, 1963, eight transistors, upper front horizontal slide rule dial with top right thumbwheel tuning, lower right thumbwheel on/off/volume knob, perforated grill area, AM, bat **$30.00**

8TP-90, vertical, 1962, eight transistors, upper front round window dial with thumbwheel tuning, lower round concentric circle grill area, AM, bat **$80.00**

9TL-365S, horizontal, 5x8½x1¾", 1962, plastic, nine transistors, upper front horizontal two-band dial with upper right thumbwheel tuning, upper left thumbwheel on/off/volume knob, two lower right switches, left metal perforated grill area, top right band switch, telescoping antenna, made in Japan, AM, SW, bat **$40.00**

9TM-40, vertical, 1961, nine transistors, shouldered case with raised top dial area, thumbwheel tuning, large lower perforated grill area, swing handle, AM, bat **$125.00**

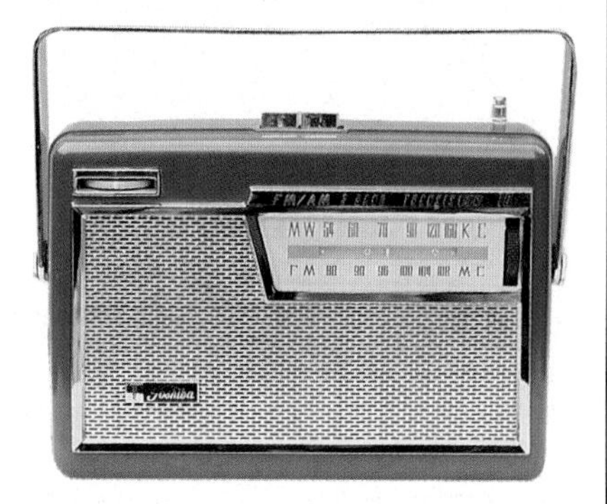

10TL-429F, horizontal, 6¼x9⅛x2⅝", 1961, plastic, 10 transistors, upper right front horizontal two-band slide rule dial with thumbwheel tuning, upper left thumbwheel on/off/volume knob, large perforated grill area, two top pushbuttons, telescoping antenna, handle, AM, FM, bat $30.00

10TL-655F, horizontal, 1963, 10 transistors, upper front horizontal two-band slide rule dial with upper right thumbwheel tuning, upper left thumbwheel on/off/volume knob, large lower perforated grill area, telescoping antenna, handle, AM, FM, bat **$30.00**

10TM-631F, horizontal, 1964, 10 transistors, two right front round dials – upper AM, lower FM – large left grill area, telescoping antenna, strap, AM, FM, bat **$30.00**

TR-193 "The Reflex," vertical, 4x2½x 1¼", 1958, plastic, four transistors,

upper right front round dial knob, left side thumbwheel on/off/volume knob, lower perforated lace grill area, AM, bat $200.00

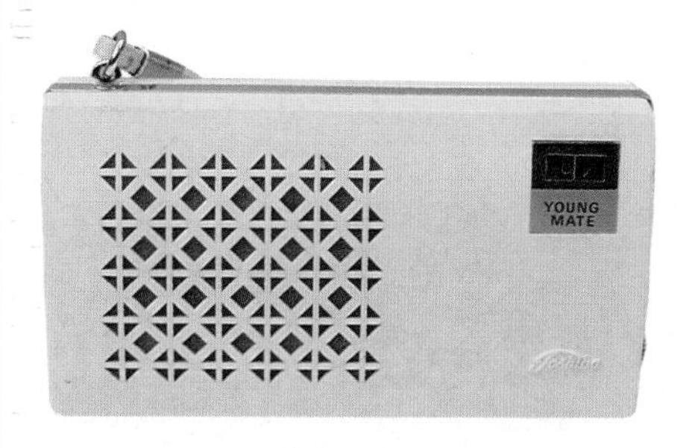

"Young Mate," horizontal, 2⅝x5x 1¼", plastic, upper right front window dial with right side thumbwheel tuning, right side thumbwheel on/off/volume knob, left grill area with criss-cross cut-outs, strap, AM, bat $20.00

Trancel

T-7, vertical, 1959, six transistors, upper front horizontal slide rule dial with thumbwheel tuning knob, left side thumbwheel on/off/volume knob, lower diagonal grill bars with lower left logo, AM, bat **$35.00**

T-11, horizontal, 1962, plastic, six transistors, right front window dial with right side thumbwheel tuning, lower right side thumbwheel on/off/volume knob, large perforated grill area, made in Japan, AM, bat $25.00

TR80, (bottom left) vertical, 4¼x2¾ x1¼", eight transistors, upper front horizontal dial with right side thumbwheel tuning, left side thumbwheel on/off/volume knob, lower metal perforated grill area, rear fold-out stand, AM, bat $35.00

TR-81, vertical, 1962, eight transistors, upper front round window dial with thumbwheel tuning, large lower perforated grill area, AM, bat **$25.00**

Trans-American

SR-6T60, vertical, 1962, six transistors, upper left front round dial knob, right side thumbwheel on/off/volume knob, lower perforated grill area, AM, bat **$30.00**

Trans-ette

6YR-21, vertical, plastic, six transistors, upper right front window dial with right side thumbwheel tuning, left side thumbwheel on/off/volume knob, lower round metal perforated grill area, AM, bat **$30.00**

TRN-3, vertical, 1961, three transistors, upper front window dial with left side thumbwheel tuning, lower perforated grill area, swing handle, AM, bat **$30.00**

YRM6, vertical, 1962, six transistors, upper left front round window dial with top thumbwheel tuning, top right thumbwheel on/off/volume knob, lower perforated grill area, AM, bat **$25.00**

Transitone

TR-1645, vertical, 1963, six transistors, step-back top, upper left window dial with thumbwheel tuning, right side thumbwheel on/off/volume knob, large grill area with horizontal bars, AM, bat **$10.00**

Transonic

9T-641, horizontal, 1964, nine transistors, two right front window dials – one AM, one FM – right side thumbwheel tuning, right side thumbwheel on/off/volume knob, left perforated grill area, telescoping antenna, AM, FM, bat **$15.00**

1095N, vertical, 1964, 10 transistors, upper left front round window dial with thumbwheel tuning, lower grill area with horizontal bars, AM, bat **$15.00**

Trav-Ler

TR-280, vertical, 1959, six transistors, upper front round dial knob, top right on/off/volume knob, lower perforated grill area, swing handle, AM, bat **$45.00**

TR-282-B, vertical, 1959, six transistors, upper front round dial knob, top right on/off/volume knob, lower metal perforated grill area with raised circular pattern, swing handle, AM, bat **$45.00**

TR-283, vertical, 1959, plastic, six transistors, upper front round dial knob, top right on/off/volume

knob, lower metal perforated grill area, swing handle, AM, bat .. $45.00

TR-601, vertical, 1962, six transistors, upper right side thumbwheel dial, upper left side thumbwheel on/off/volume knob, lower perforated grill area with lower left logo, AM, bat **$35.00**

TR-630, horizontal, 1962, nine transistors, right front dial with right side thumbwheel tuning, lower right side thumbwheel on/off/volume knob, left lattice grill area with upper left logo, AM, bat **$30.00**

Trend

FR-625PH "Super Deluxe," vertical, 1965, six transistors, upper right front window dial with right side thumbwheel tuning, left side thumbwheel on/off/volume knob, lower grill area, AM, bat **$15.00**

Truetone

3326, horizontal watch/radio, 1964, six transistors, upper left front horizontal slide rule dial, three thumbwheel knobs, right round watch face overlaps lower perforated grill area, swing handle, AM, bat **$60.00**

D3614A, horizontal, 1957, plastic, four transistors, right front round dial knob, lower thumbwheel on/off/volume knob, center checkered grill area, right side strap, AM, bat **$150.00**

D3715A, horizontal, 1958, plastic, four transistors, right front round dial knob, lower on/off/volume knob, center checkered grill area, AM, bat **$125.00**

D3716A, horizontal, 1957, leather, five transistors, upper right front dial knob, upper left front on/off/volume knob, lower grill area with rectangular cut-outs, leather handle, AM, bat **$30.00**

D3716B, horizontal, 1957, leather, five transistors, upper right front dial knob, upper left front on/off/volume knob, lower grill area with rectangular cut-outs, leather handle, AM, bat **$30.00**

DC1400, horizontal/clock radio, 1964, six transistors, wedge-shaped case, right front dial with upper right thumbwheel tuning, lower right thumbwheel on/off/volume knob, large left grill area, AM, bat **$15.00**

DC3050, horizontal, 1959, eight transistors, two upper front horizontal slide rule dials, upper right thumbwheel tuning, upper left thumbwheel on/off/volume knob, lower perforated grill area, telescoping antenna, AM, SW, bat **$30.00**

DC3052, horizontal, 2⅝x4½x1⅜", 1959, plastic, six transistors, right front V-shaped window dial over large perforated grill area, right side thumbwheel tuning, lower right side thumbwheel on/off/volume knob, AM, bat **$40.00**

DC3084A, horizontal, 1960, leather, six transistors, upper right front dial knob, upper left front on/off/volume knob, center grill area with rectangular cut-outs, leather handle, AM, bat **$20.00**

DC3088A, horizontal, 1960, eight transistors, upper right front dial knob, lower right on/off/volume knob over large perforated grill area with lower left logo, fold-down handle, AM, bat **$20.00**

DC3090, vertical, 1960, three transistors, upper front window dial with thumbwheel tuning, lower round perforated grill area, swing handle, AM, bat **$50.00**

DC3164A, vertical, 1962, six transistors, upper front round window dial with right side thumbwheel tuning, right side thumbwheel on/off/volume knob, lower perforated grill area, AM, bat **$20.00**

DC3280, horizontal, 1962, eight transistors, upper front horizontal two-band dial with right thumbwheel tuning, top left thumbwheel on/off/ volume knob, lower perforated grill area, AM, SW, bat **$25.00**

DC3306, horizontal, 1963, six transistors, upper right front window dial with right side thumbwheel tuning, lower right side thumbwheel on/off/ volume knob, oval perforated grill area, AM, bat **$20.00**

DC3316, vertical, 1963, six transistors, upper front horizontal slide rule dial with thumbwheel tuning, large lower perforated grill area, AM, bat **$15.00**

DC3318, vertical, 1963, eight transistors, upper front horizontal slide rule dial with thumbwheel tuning, lower perforated grill area, AM, bat **$15.00**

DC3338, vertical, 1963, eight transistors, two upper right front windows – one for stations, one for CD marks – lower perforated grill area, strap, AM, bat **$20.00**

DC3350, horizontal, 1964, 10 transistors, upper front horizontal two-band dial, right and left side knobs, large lower perforated grill area with lower right and left knobs, telescoping antenna, fold-down handle, AM, FM, bat **$20.00**

DC3406, vertical, 1963, six transistors, step-back top, upper left front window dial with thumbwheel tuning, right side thumbwheel on/off/volume knob, large grill area with horizontal bars, AM, bat **$10.00**

DC3407, vertical, 1964, seven transistors, upper front round dial knob, right side thumbwheel on/ off/volume knob, lower lattice grill area with lower left logo, AM, bat **$15.00**

DC3408, vertical, 4¾x3x1¼", 1964, plastic, eight transistors, upper front horizontal slide rule dial with right side thumbwheel tuning, right side thumbwheel on/off/volume knob, lower oval perforated grill area with center logo, AM, bat **$20.00**

DC3408B, vertical, 4¾x3x1¼", 1964, plastic, eight transistors, upper front horizontal slide rule dial with right side thumbwheel tuning, right side thumbwheel on/off/volume knob, lower oval perforated grill area with center logo, AM, bat **$20.00**

DC3416, vertical, 1964, six transistors, upper front horizontal slide rule dial with thumbwheel tuning, large lower perforated grill area, AM, bat **$10.00**

DC3426, horizontal watch/radio, 1964, six transistors, upper left front horizontal slide rule dial, three thumbwheel knobs, right round watch face overlaps lower perforated grill area, swing handle, AM, bat **$60.00**

DC3429B, horizontal, 1964, nine transistors, right front dial with thumbwheel tuning, upper left on/ off/volume knob, large checkered grill area, handle, AM, bat..... **$15.00**

DC3436, horizontal, 1964, six transistors, left front vertical slide rule dial with top left thumbwheel tuning, top right thumbwheel on/off/volume knob, oval perforated grill area with center logo, AM, bat **$20.00**

DC3438, vertical, 1963, eight transistors, two upper right front windows – one for stations, one for CD marks – lower perforated grill area, strap, AM, bat **$20.00**

DC3448, horizontal, 1963, nine transistors, upper front horizontal three-band dial, thumbwheel knobs, lower perforated grill area, swing handle, AM, 2SW, bat **$30.00**

DC3459, horizontal, 1963, 10 transistors, upper front horizontal two-band slide rule dial with right tuning knob, top left thumbwheel knob, large lower perforated grill area, telescoping antenna, AM, FM, bat **$20.00**

DC3460, horizontal, 1964, 10 transistors, upper front horizontal two-band dial, right and left side knobs, large lower perforated grill area with lower right and left knobs, telescoping antenna, fold-down handle, AM, FM, bat...**$20.00**

DC3506, horizontal, 1964, six transistors, upper right front window dial with right side thumbwheel tuning, lower right side thumbwheel on/off/volume knob, oval perforated grill area, AM, bat **$20.00**

DC3609 "Hi Fidelity," horizontal, 1965, leather, nine transistors, upper front horizontal slide rule dial, right dial and Hi/Lo switch, left perforated grill area, leather handle, AM, bat **$15.00**

DC3610, horizontal, 1965, 10 transistors, right front window dial with upper right side thumbwheel tuning, lower right side thumbwheel on/off/volume knob, large perforated grill area, AM, bat **$15.00**

DC3612, vertical, 1965, 12 transistors, upper front round dial knob, right thumbwheel on/off/volume knob, lower perforated grill area, AM, bat **$15.00**

DC3654, horizontal, 1965, 10 transistors, upper front horizontal two-band slide rule dial with right knob, lower grill area with vertical bars and lower right FM/AM switch, telescoping antenna, AM, FM, bat **$15.00**

DC3704 "Jr.," vertical, 3⅞x2½x1¼", plastic, six transistors, upper right front window dial with right side thumbwheel tuning, left side thumbwheel on/off/volume knob, lower grill area with horizontal bars, made in Hong Kong, AM, bat $15.00

DC3884, horizontal, 1959, leather, four transistors, upper right front dial knob, upper left front volume knob, center grill area with circular cut-outs, leather handle, AM, bat **$25.00**

DC3886A, horizontal, 1958, leather, six transistors, upper right front dial knob, upper left front volume knob, center grill area with circular cut-outs, leather handle, AM, bat **$25.00**

United Royal

801-T, horizontal, 1962, eight transistors, upper right front window dial with thumbwheel tuning, large perforated grill area with lower left logo, AM, bat **$25.00**

802, horizontal, 3⅝x5⅞x1½", plastic, eight transistors, upper front horizontal two-band dial with right side thumbwheel tuning, right side thumbwheel on/off/volume knob, right side BC/SW switch, large metal perforated grill area, made in Japan, AM, SW, bat $20.00

1050, horizontal, plastic, top raised two-band horizontal slide rule dial, top right tuning knob, top left volume/tone knob, large front perforated grill area with lower right logo, three pushbuttons, telescoping antenna, handle, AM, FM, bat ... **$30.00**

"Ten Hundred," horizontal, 3½x6x 1½", 1964, plastic, nine transistors, upper right front window dial with right side thumbwheel tuning, right side thumbwheel on/off/volume knob, large metal perforated grill area with lower left logo, made in Japan, AM, bat $30.00

Universal

PTR-62B, vertical, 4¼x2¾x1¼", 1963, plastic, six transistors, upper right front thumbwheel dial, upper

left front thumbwheel on/off/volume knob, lower metal perforated grill area, made in Japan, AM, bat $25.00

PTR-81B, vertical, 1962, plastic, upper right front thumbwheel dial, upper left front thumbwheel on/off/volume knob, lower metal perforated grill area with lower left logo, AM, bat **$25.00**

RE-64 "Deluxe," vertical, 4x2⅝x1", 1964, plastic, six transistors, upper right front window dial with right side thumbwheel tuning, left side thumbwheel on/off/volume knob, lattice grill area, made in Taiwan, AM, bat $10.00

YT-161, vertical, 4⅛x2½x1⅜", plastic, six transistors, upper right thumbwheel dial knob, left on/off/volume window with left side thumbwheel knob, lower metal perforated grill area, made in Japan, AM, bat **$25.00**

Valiant

AM1400 "Hi Power," vertical, upper left front round dial knob, upper right on/off/volume knob, lower grill area with vertical slots, AM, bat **$30.00**

HT-1200 "High Fidelity," vertical, 4¼x2⅝x1¼", plastic, 10 transistors, upper right front triangular window dial with right side thumbwheel tuning, top left thumbwheel on/off/volume knob, lower metal perforated grill area with criss-cross pattern, made in Japan, AM, bat **$25.00**

HT-6043 "Deluxe HiFi," vertical, 4¼x2½x1¼", plastic, six transistors, upper right front square window dial with right side thumbwheel tuning, top left thumbwheel on/off/volume knob, lower metal perforated grill area, AM, bat $20.00

V-666-A "DeLuxe," vertical, 4¼x2⅝x 1¼", plastic, six transistors, upper right front window dial with right side thumbwheel tuning, left side thumbwheel on/off/volume knob, vertical grill bars, made in Hong Kong, AM, bat $10.00

Vesper

G-1110, horizontal, 1963, nine transistors, right front vertical two-band slide rule dial with thumbwheel tuning, lower FM/AM switch, left perforated grill area, telescoping antenna, AM, FM, bat **$15.00**

Victoria

TR-650, vertical, 1961, six transistors, upper front horizontal dial with thumbwheel tuning, lower perforated grill area, AM, bat **$20.00**

Viscount

6TP-103, vertical, 4¼x2¾x1¼", plastic, six transistors, upper front see-through window dial with thumbwheel tuning, upper right thumbwheel on/off/volume knob, lower metal perforated grill area, AM, bat $35.00

601, vertical, 1965, six transistors, upper right front window dial with right side thumbwheel tuning, left side thumbwheel on/off/volume knob, lower horizontal grill bars, strap, AM, bat **$10.00**

602 "VIP," vertical, 1962, plastic, six transistors, upper front wedge-shaped see-through window dial with thumbwheel tuning, upper right thumbwheel on/off/volume knob, lower metal perforated grill area, AM, bat **$40.00**

616, vertical, plastic, six transistors, upper right front window dial with right side thumbwheel tuning, top left thumbwheel on/off/volume knob, lower metal perforated grill area, AM, bat $20.00

712, vertical, 1965, seven transistors, upper front round dial knob, right side thumbwheel on/off/volume knob, lower horizontal grill bars, AM, bat.................................. **$10.00**

815, vertical, 4¼x2½x1½", plastic, eight transistors, upper front round dial knob, right side thumbwheel on/off/volume knob, lower horizontal grill bars, made in Hong Kong, AM, bat.................................. **$10.00**

833, horizontal, 4¾x8½x1¾", plastic, eight transistors, upper front horizontal three-band slide rule dial with right tuning, upper left thumbwheel on/off/volume knob, top

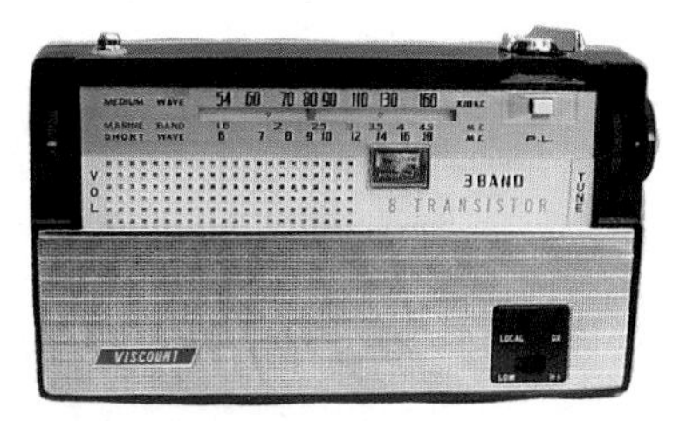

right switch, lower metal perforated grill area with two switches, AM, SW, Marine, bat $35.00

1022, vertical, 1965, 10 transistors, upper right front window dial with right side thumbwheel tuning, lower perforated grill area, AM, bat..................... **$10.00**

1030, horizontal, 3½x6⅜x1⅛", plastic, 10 transistors, large right front window dial with right thumbwheel tuning, top left thumbwheel on/off/volume knob, left metal perforated grill area, made in Japan, AM, bat................ $25.00

Vista

10, horizontal, 1965, 10 transistors, right front vertical two-band slide rule dial, large left perforated grill area, telescoping antenna, strap, AM, FM, bat **$15.00**

G-1050, horizontal, 1964, 10 transistors, right front two-band thumbwheel dial, upper AM/FM switch, upper left thumbwheel on/off/volume knob, left grill area with horizontal slots, telescoping antenna, AM, FM, bat $15.00

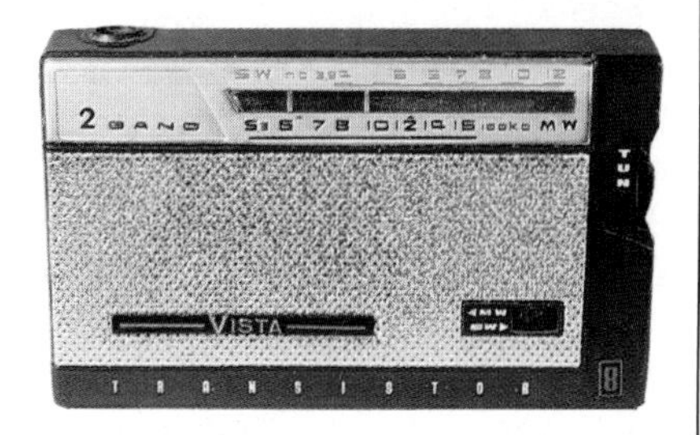

NTR-800, horizontal, 2⅞x4⅞x1¼", 1964, plastic, eight transistors, upper front horizontal two-band slide rule dial with right side thumbwheel tuning, right side thumbwheel on/off/ volume knob, metal perforated grill area with MW/SW switch, telescoping antenna, made in Japan, AM, SW, bat $25.00

NTR-850, horizontal, 1963, eight transistors, upper front horizontal slide rule dial with right side thumbwheel tuning, oval perforated grill area, AM, bat $25.00

NTR-966, vertical, 1964, upper front window dial with right side thumbwheel tuning, lower round perforated grill area, AM, bat $30.00

Vornado

V-700, vertical, 1965, eight transistors, upper front window dial with right side thumbwheel tuning, large lower tear drop-shaped perforated grill area, AM, bat $40.00

V-820, vertical, 1965, eight transistors, upper front horizontal slide rule dial with right side thumbwheel tuning, lower perforated grill area, AM, bat $20.00

V-1670, vertical, 1965, six transistors, upper right front window dial with right side thumbwheel tuning, lower grill area with horizontal bars, AM, bat $20.00

V-2091, horizontal, 1965, 10 transistors, upper front horizontal two-band slide rule dial, large lower perforated grill area with lower right FM/AM switch, AM, FM, bat $15.00

Vulcan

6T-160, horizontal, 2¾x4½x1¼", 1960, plastic, six transistors, upper right front thumbwheel dial behind see-through panel, upper left thumbwheel on/off/volume knob, lower metal perforated grill area, made in Japan, AM, bat $25.00

Waltham

WA1001, vertical, 1965, six transistors, upper right front window dial with right side thumbwheel tuning, lower grill area, right side strap, AM, bat $10.00

WA3201L, horizontal, 1965, leather, eight transistors, right front vertical slide rule dial with right side tuning, right side on/off/volume knob, left perforated grill area, leather handle, AM, bat **$15.00**

WA5001, vertical, 1965, 10 transistors, upper right front window dial with right side thumbwheel tuning, top left thumbwheel on/off/volume knob, left vertical perforated grill area, AM, bat **$15.00**

WA5102L, horizontal, 1965, leather, 10 transistors, upper right front dial knob, lower right thumbwheel on/off/volume knob, left perforated grill area with lower left logo, handle, AM, bat **$20.00**

WB5201L, horizontal, 1965, leather, 10 transistors, right front vertical two-band slide rule dial with right side tuning, right side on/off/volume knob, left perforated grill area, leather handle, AM, FM, bat... **$15.00**

WA5303L, horizontal, 1965, leather, 10 transistors, upper front horizontal slide rule dial, three right knobs, left grill area with horizontal bars, leather handle, AM, bat **$15.00**

WB4101, horizontal, 1965, nine transistors, upper right front round two-band dial, right side thumbwheel knobs, left grill area with horizontal bars, telescoping antenna, strap, AM, FM, bat **$15.00**

WB7303L, horizontal, 1965, 12 transistors, upper front horizontal two-band slide rule dial, four right knobs, left grill area with horizontal bars, telescoping antenna, leather handle, AM, FM, bat **$15.00**

Watterson

601, horizontal/table, 1958, wood, six transistors, right front dial, three knobs, left grill area, feet, AM, bat **$20.00**

Webcor

310, horizontal, 1961, 10 transistors, off-center vertical two-band slide rule dial over large perforated grill area with lower left logo, top thumbwheel tuning and on/off/volume knobs, telescoping antenna, AM, FM, bat **$25.00**

B-308, horizontal, 1962, eight transistors, upper front horizontal slide rule dial with top thumbwheel tuning, lower right thumbwheel on/off/volume knob, perforated grill area, telescoping antenna, AM, bat **$35.00**

E306, horizontal, 1961, seven transistors, upper front horizontal two-band slide rule dial with top thumbwheel tuning, lower right thumbwheel on/off/volume knob, perforated grill area, telescoping antenna, AM, SW, bat **$35.00**

E307, vertical, 1964, six transistors, upper front slanted horizontal slide rule dial with thumbwheel tuning, large lower perforated grill area with center logo, AM, bat **$20.00**

E-312, horizontal, 1962, eight transistors, upper front horizontal three-band slide rule dial, lower left grill area with horizontal slots, telescoping antenna, AM, SW, LW, bat **$20.00**

E313, horizontal, 1964, six transistors, upper right front round window dial with right side thumbwheel tuning, right side thumbwheel on/off/volume knob, large left perforated grill area, AM, bat.......................... **$15.00**

E314, vertical, 1964, eight transistors, upper front horizontal slide rule dial with right side thumbwheel tuning, right side thumbwheel on/off/volume knob, perforated grill area with lower left logo, AM, bat **$15.00**

E316 "Patio," horizontal, 1964, nine transistors, upper front horizontal two-band slide rule dial, four right knobs, large lower checkered grill area, telescoping antenna, handle, AM, FM, bat **$15.00**

G-309, horizontal, 1962, eight transistors, upper front horizontal two-band slide rule dial with right side thumbwheel tuning, top left thumbwheel on/off/volume knob, perforated grill area, battery window, AM, Weather, bat **$35.00**

R305, vertical, 1961, six transistors, upper left front window dial with left side thumbwheel tuning, upper right thumbwheel on/off/volume knob, lower perforated grill area, swing handle, AM, bat **$30.00**

Westclox

80002, horizontal/clock radio, 1962, six transistors, upper right front window dial over perforated grill area, right side thumbwheel tuning and on/off/volume knobs, left front clock face, AM, bat **$25.00**

80004, horizontal/clock radio, 1962, six transistors, upper right front window dial over perforated grill area, right side thumbwheel tuning and on/off/volume knobs, left front clock face, AM, bat **$25.00**

80006, horizontal/clock radio, 1962, six transistors, upper right front window dial over perforated grill area, right side thumbwheel tuning and on/off/volume knobs, left front clock face, AM, bat **$25.00**

Westinghouse

H-602P7, horizontal, 1957, leather, seven transistors, upper right front dial knob, upper left front on/off/volume knob, lower metal perforated grill area, leather handle, AM, bat **$30.00**

H610P5, horizontal, 1957, charcoal gray, five transistors, large right front round dial knob, lower thumbwheel on/off/volume knob, left checkered grill area, AM, bat **$75.00**

H611P5, horizontal, 1957, blue, five transistors, large right front round dial knob, lower thumbwheel on/off/volume knob, left checkered grill area, AM, bat **$75.00**

H612P5, horizontal, 1957, yellow, five transistors, large right front round dial knob, lower thumbwheel on/off/volume knob, left checkered grill area, AM, bat **$75.00**

H-617P7, horizontal, 1957, seven transistors, large right front round dial knob, lower thumbwheel on/off/volume knob, left checkered grill area, AM, bat **$40.00**

H621P6, horizontal, 6½x9¼x3½", 1958, charcoal plastic, six transistors, right front round dial knob, left front round on/off/volume knob, top raised lattice grill area, swing handle, AM, bat **$45.00**

H622P6, horizontal, 6½x9¼x3½", 1958, yellow/white plastic, six transistors, right front round dial knob, left front round on/off/volume knob, top raised lattice grill area, swing handle, AM, bat **$45.00**

H-652P6, horizontal, 3x6x1½", 1958, plastic, six transistors, right front round dial knob overlaps top checkered grill area, lower left perforated grill area, lower right thumbwheel on/off/volume knob, AM, bat **$30.00**

H-653P6, horizontal, 3x6x1½", 1958, plastic, six transistors, right front round dial knob overlaps top checkered grill area, lower left perforated grill area, lower right thumbwheel on/off/volume knob, AM, bat $30.00

H655P5, vertical, 1959, plastic, five transistors, lower front round dial knob overlaps upper checkered grill area, right side on/off/volume knob, swing handle, AM, bat **$30.00**

H656P5, vertical, 1959, plastic, five transistors, lower front round dial knob overlaps upper checkered grill area, right side on/off/volume knob, swing handle, AM, bat **$30.00**

H-685P8, horizontal/clock radio, 4⅛x9⅝x3", 1959, white/brown plastic, eight transistors, right front round dial over large perforated grill area, lower thumbwheel alarm/volume knob, left alarm clock face, feet, AM, bat **$20.00**

H-686P8, horizontal/clock radio, 4⅛x9⅝x3", 1959, white/pink plastic, eight transistors, right front round dial over large perforated grill area, lower thumbwheel alarm/volume knob, left alarm clock face, feet, AM, bat **$20.00**

H-690P5, horizontal, 6⅝x7¾x2¾", plastic, five transistors, large right front round dial knob overlaps grill area with circular cut-outs, left side on/off/volume knob, handle, AM, bat **$25.00**

H-693P8, horizontal, 3⅛x6x2", 1959, brown plastic, eight transistors, right front round convex dial knob overlaps top checkered grill area, lower left perforated grill area, lower right thumbwheel on/off/volume knob, AM, bat **$35.00**

H-694P8, horizontal, 3⅛x6x2", 1959, green plastic, eight transistors, right front round convex dial knob overlaps top checkered grill area, lower left perforated grill area, lower right thumbwheel on/off/volume knob, AM, bat $35.00

H-695P8, horizontal, 3⅛x6x2", 1959, pink plastic, eight transistors, right front round convex dial knob overlaps top checkered grill area, lower left perforated grill area, lower right thumbwheel on/off/volume knob, AM, bat $35.00

H-697P7, vertical, 7x4¼x2⅛", 1959, charcoal gray/white plastic, seven transistors, lower front round dial knob overlaps upper checkered grill area, right side on/off/volume knob, swing handle, AM, bat **$30.00**

H-698P7, vertical, 7x4¼x2⅛", 1959, yellow/white plastic, seven transistors, lower front round dial knob overlaps upper checkered grill area, right side on/off/volume knob, swing handle, AM, bat **$30.00**

H-699P7, vertical, 7x4¼x2⅛", 1959, green/white plastic, seven transistors, lower front round dial knob overlaps upper checkered grill area, right side on/off/volume knob, swing handle, AM, bat $30.00

H-707P6GP, vertical, 3¾x2½x1⅛", green plastic, six transistors, upper right front window dial with right side thumbwheel tuning, top left thumbwheel on/off/volume knob, lower metal perforated grill area with lower left logo, made in Japan, AM, bat **$20.00**

H-713P9, horizontal, 1960, nine transistors, two upper front horizontal slide rule dial scales, large lower grill area, telescoping antenna, handle, AM, SW, bat **$20.00**

H-725P6A, horizontal, 1961, white/red plastic, six transistors, large right front round dial knob overlaps grill area with circular cut-outs, left side on/off/volume knob, handle, AM, bat **$25.00**

H-726P6, horizontal, 1960, white/turquoise plastic, six transistors, large right front round dial knob overlaps grill area with circular cut-outs, left side on/off/volume knob, handle, AM, bat **$25.00**

H-730P7, horizontal, 1960, leather, seven transistors, right front round dial with right side tuning knob, left side on/off/volume knob, left grill area, leather handle, AM, bat **$15.00**

H-733P7, vertical, 1961, seven transistors, upper left front round window dial with thumbwheel tuning, lower perforated grill area, AM, bat **$15.00**

H-737P7, horizontal, 1961, seven transistors, upper front horizontal slide rule dial with right side tuning, left side on/off/volume knob, lower grill area with horizontal bars, AM, bat **$15.00**

H769P7A, horizontal, 1960, tan leather, seven transistors, right front round dial with right side tuning knob, left side on/off/volume knob, left checkered grill area, leather handle, AM, bat **$15.00**

H770P7A, horizontal, 1960, gray leather, seven transistors, right front round dial with right side tuning knob, left side on/off/volume knob, left checkered grill area, leather handle, AM, bat **$15.00**

H-772P6GP, horizontal, 1961, plastic, six transistors, large right front round dial knob overlaps grill area with circular cut-outs, left side on/off/volume knob, handle, AM, bat **$25.00**

H-790P6, vertical, 7x4¼x2", 1962, plastic, six transistors, lower front round dial overlaps upper horizontal grill bars, lower right side on/off/volume knob, swing handle, AM, bat **$20.00**

H-790P6GP, vertical, 7x4¼x2", 1962, plastic, six transistors, lower front round dial overlaps upper horizontal grill bars, lower right side on/off/volume knob, swing handle, AM, bat $20.00

H-791P6, vertical, 7x4¼x2", 1962, plastic, six transistors, lower front round dial overlaps upper horizontal grill bars, lower right side on/off/volume knob, swing handle, AM, bat $20.00

H-791P6GP, vertical, 7x4¼x2", 1962, plastic, six transistors, lower front round dial overlaps upper horizontal grill bars, lower right side on/off/volume knob, swing handle, AM, bat $20.00

H-793P6GP, horizontal, 1962, white/charcoal plastic, six transistors, large right front round dial knob overlaps grill area with circular cut-outs, left side on/off/volume knob, handle, AM, bat $20.00

H-796P6, vertical, 4½x2½x1½", 1962, six transistors, upper left front round window dial with thumb-

wheel tuning, lower round perforated grill area, AM, bat $15.00

H-812P8, horizontal, 1963, eight transistors, flip-up top with cut-outs for right and left top knobs, inner three-band slide rule dial, front perforated grill area with lower right logo, telescoping antenna, strap, AM, 2SW, bat $40.00

H-842P6, horizontal, 2½x5x1", 1963, charcoal plastic, six transistors, large right front round dial knob, top left thumbwheel on/off/volume knob, left grill area with horizontal bars, AM, bat $15.00

H-868P12, horizontal, $7x9\frac{1}{2}x3\frac{1}{2}$", 1963, leather, 12 transistors, two upper front pointer dials – right AM, left FM – center band switch, lower right thumbwheel knobs, lower left perforated grill area, telescoping antenna, handle, AM, FM, bat ... **$20.00**

H-901P7GP, vertical, 1964, seven transistors, large upper left front round dial knob overlaps lower grill area with horizontal bars and lower right logo, AM, bat **$15.00**

H-902P6GP, vertical, 1965, plastic, six transistors, upper right front window dial with right side thumbwheel tuning, top left thumbwheel on/off/volume knob, lower perforated grill area with lower left logo, AM, bat **$15.00**

H-902P6GPA, vertical, $3\frac{3}{4}x2\frac{3}{8}x1$", 1967, plastic, six transistors, upper right front window dial with right side thumbwheel tuning, top left thumbwheel on/off/volume knob, lower metal perforated grill area with lower left logo, made in Japan, AM, bat $15.00

H-908PN9GP, vertical, 1965, nine transistors, two upper front horizontal slide rule dial scales – one FM, one AM – right side thumbwheel knobs, lower perforated grill area with lower right logo, telescoping antenna, AM, FM, bat **$15.00**

H-914P8GP, vertical, $4\frac{1}{4}x2\frac{5}{8}x1\frac{1}{4}$", 1965, black plastic, eight transistors, upper right front window dial with right side thumbwheel tuning, upper left front on/off/volume window with left side thumbwheel knob, lower metal perforated grill area, made in Japan, AM, bat $15.00

H-939P8GP, vertical, $4\frac{1}{4}x2\frac{5}{8}x1\frac{1}{4}$", white plastic, eight transistors, upper right front window dial with right side thumbwheel tuning, upper left front on/off/volume window with left side thumbwheel knob, lower metal perforated grill area, made in Japan, AM, bat **$15.00**

RS11P28-A "Escort," horizontal/radio flashlight, 3¼x4¼x1¼", plastic, upper right front window dial, lower metal perforated grill area, left side flashlight, built-in recharger, AM, bat **$20.00**

RS21P08A "Escort," horizontal/radio flashlight-watch-cigarette lighter, 3¼x4¼x1¼", plastic, upper right front window dial, upper left watch face, lower metal perforated grill area, left side flashlight and cigarette lighter, built-in recharger, AM, bat $40.00

Wilco

ST-6, horizontal, 1962, six transistors, right front V-shaped window dial with right side thumbwheel tuning, lower right round on/off/volume window with right side thumbwheel knob, left perforated grill area with upper left logo, AM, bat **$30.00**

ST-88, horizontal, 1963, eight transistors, right front dial with right side thumbwheel tuning, lower right round on/off/volume window with right side thumbwheel knob, left perforated grill area with upper left logo, AM, bat **$20.00**

Windsor

"Coronet Boy's Radio," vertical, 4x2½x1¼", plastic, two transistors, upper right front "crown" window dial with right side thumbwheel tuning, left front on/off/volume window with left side thumbwheel knob, metal perforated grill area with Windsor logo, telescoping antenna, made in Japan, AM, bat $40.00

Winston

W111, vertical, 1965, 11 transistors, upper left front window dial with thumbwheel tuning, large lower perforated grill area, AM, bat **$15.00**

W700, vertical, 1965, seven transistors, right side tuning and on/off/volume knobs, front perforated grill area, left strap, AM, bat **$30.00**

Yaecon

YTR-58, vertical, 4x2½x1½", 1960, plastic, six transistors, upper left front window dial with left side thumbwheel tuning, right side thumbwheel on/off/volume knob, lower metal perforated grill area with lower right logo, made in Japan, AM, bat $40.00

YTR-721, horizontal, plastic, seven transistors, two right front square window dials – one BC, one SW – with right side thumbwheel tuning, lower right front round on/off/volume window with right side thumbwheel

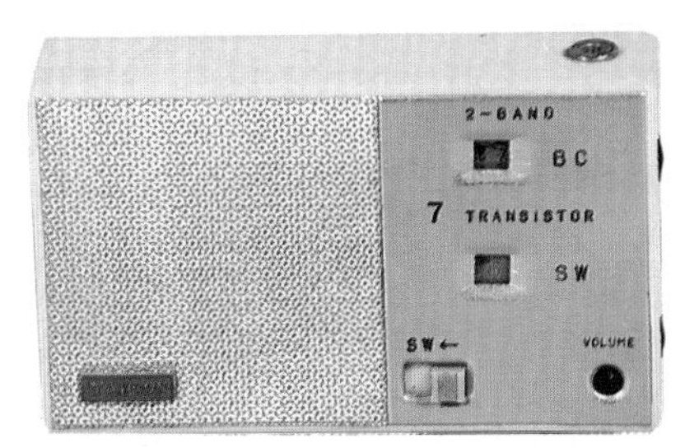

knob, band switch, left metal perforated grill area with lower left logo, BC, SW, bat $30.00

Yashica

YT-100, vertical, 1961, six transistors, upper right front round dial knob overlaps large lower perforated grill area with vertical lines and lower left logo, swing handle, AM, bat ... **$25.00**

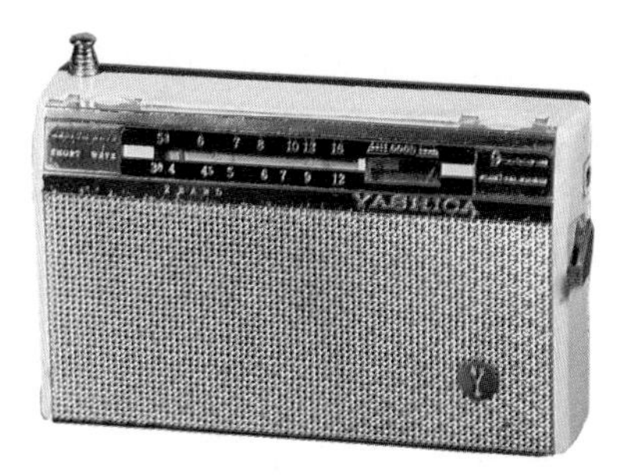

YT-300, horizontal, 1961, plastic, nine transistors, upper front horizontal two-band slide rule dial with top right thumbwheel tuning, top left thumbwheel on/off/volume knob, battery window, large lower perforated grill area with lower right logo, right side switch, telescoping antenna, AM, SW, bat $25.00

York

TR-65, vertical, 1965, six transistors, upper right front window dial with right side thumbwheel tuning, lower textured grill area, AM, bat ... **$10.00**

TR-89, vertical, 1965, eight transistors, upper right front window dial with right side thumbwheel tuning, lower patterned grill area, AM, bat **$10.00**

TR-100, horizontal, 1963, leather, 10 transistors, right front thumbwheel dial, top left thumbwheel on/off/volume knob, large grill area with horizontal slots, handle, AM, bat **$15.00**

TR-101, horizontal, 1965, 10 transistors, right front window dial with right side thumbwheel tuning, large left perforated grill area, AM, bat **$15.00**

TR-103, vertical, 1965, 10 transistors, large round upper front dial knob, upper right thumbwheel on/off/volume knob, lower grill area, AM, bat **$15.00**

TR-105, horizontal, 1965, 10 transistors, right front two-band thumbwheel dial, lower on/off/volume knob, upper FM/AM switch, large left grill area, telescoping antenna, AM, FM, bat **$15.00**

TR-121, horizontal, 1965, leather, 12 transistors, right front two-band thumbwheel dial, upper FM/AM switch, upper left thumbwheel on/off/volume knob, round perforated grill area, telescoping antenna, AM, FM, bat **$20.00**

TR-122, vertical, 1965, 12 transistors, large round upper front dial knob, upper right thumbwheel on/off/volume knob, lower perforated grill area, AM, bat **$15.00**

TR-123, horizontal, 1965, leather, 12 transistors, right front two-band thumbwheel dial, lower on/off/volume knob, upper FM/AM switch, large left perforated grill area, telescoping antenna, handle, AM, FM, bat **$15.00**

Zenith

Zenith "Royal" model numbers can be confusing at times. For example, the same set that is marked on the front of the case "Royal 41" can have a paper label inside the case with a model number "R41W." Because most transistor collectors use the "Royal" number from the front of the case for identification purposes (rather than the model number from the inside label), to keep Zenith identification as simple as possible we have decided to use the common practice of referring to the "Royal" number from the case front when listing the following sets.

Royal 16, vertical/billfold-style, 5⅜x 3½x1⅜" (closed), plastic outer case with perforated front cover, inner

metal perforated grill area with upper right window dial, two right side thumbwheel knobs, made in Japan, AM, bat $25.00

Royal 20, vertical, 3x2⅜x1¼", plastic, eight transistors, upper front window dial with top thumbwheel tuning, left side thumbwheel on/off/volume knob, chrome front grill area with vertical bars, made in Hong Kong, AM, bat $30.00

Royal 40, vertical, 4¼x2¾x1¼", 1963, plastic, large upper front round dial knob with center pointer arrow, right side thumbwheel on/off/volume knob, lower grill area with vertical bars, AM, bat **$25.00**

Royal 41, horizontal, 4⅝x5¼x1¾", 1966, plastic, upper front horizontal dial with sliding pointer, right side thumbwheel tuning knob, left side thumbwheel on/off/volume knob, lower metal grill area with horizontal bars, AM, bat **$30.00**

Royal 50, vertical/several variations, 4¼x2¾x1½", 1962, plastic, upper front round dial knob overlaps lower grill area with vertical slots, right side thumbwheel on/off/volume knob, made in USA, AM, bat $30.00

Royal 56 "Sun Charger," horizontal, 4⅝x5¼x1¾", 1966, plastic, upper front horizontal dial with sliding pointer, right side thumbwheel tuning, left side thumbwheel on/off/volume knob, lower metal grill area with horizontal bars, swing handle with built-in solar panel, AM, bat **$125.00**

Royal 59, vertical, 4½x2¾x1⅜", plastic, eight transistors, upper front round dial with top thumbwheel tuning, right side thumbwheel on/off/volume knob, lower metal perforated grill area, AM, bat **$30.00**

Royal 60, vertical, 4⅜x2¾x1½", plastic, upper front round dial knob over large metal perforated grill area, right side thumbwheel on/off/volume knob, made in USA, AM, bat $30.00

Royal 70, horizontal, leather, upper front horizontal slide rule dial, right and left knobs overlap large lower metal perforated grill area with lower right logo and lower left tone switch, leather handle, AM, bat $20.00

Royal 74, horizontal, 6½x9½x3¾", leather, upper front horizontal slide rule dial, lower metal checkered grill area with right and left knobs, lower right tone switch, pull-up handle, AM, bat $25.00

Royal 80, vertical, 4⅜x2½x1⅜", plastic, eight transistors, recessed right panel with thumbwheel tuning and on/off/volume knobs, large front

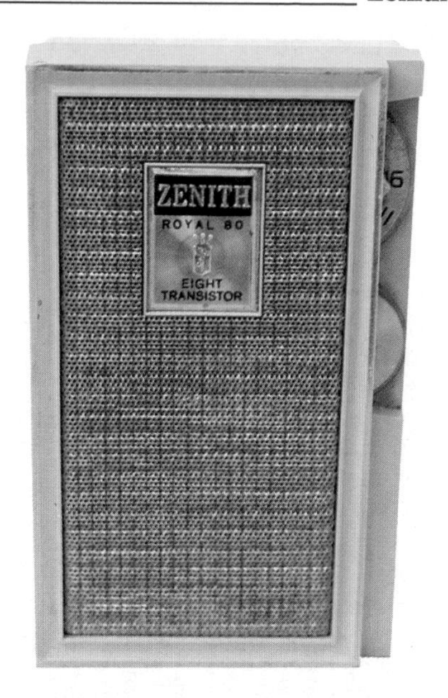

textured/perforated grill area with logo, AM, bat $40.00

Royal 85, vertical, 4⅜x2⅞x1½", plastic, eight transistors, upper front dial with semi-circular pointer and right

side thumbwheel tuning, right side thumbwheel on/off/volume knob, lower metal perforated grill area with lower left logo, AM, bat $25.00

Royal 90, vertical, upper front round dial knob overlaps large perforated grill area, right side thumbwheel on/off/volume knob, AM, bat..... **$40.00**

Royal 100 "Zenette," vertical, 1961, six transistors, upper right front round dial knob overlaps large lower criss-cross grill area, rear fold-out stand, AM, bat **$35.00**

Royal 150, vertical, 1962, six transistors, upper right front round dial knob over large perforated grill area, lower left logo, AM, bat **$30.00**

Royal 185, vertical, plastic, left front vertical slide rule dial over large metal perforated grill area with upper right logo, two right side thumbwheel knobs, AM, bat $30.00

Royal 200, vertical, 1959, plastic, seven transistors, upper front round dial knob, lower on/off/volume knob over checkered grill area with lower right logo, swing handle, AM, bat $45.00

Royal 250, vertical, 5$\frac{7}{8}$x3$\frac{3}{4}$x1$\frac{3}{4}$", 1959, plastic, six transistors, upper right front dial knob, upper left front on/off/volume knob, left grill area with horizontal bars, swing handle, made in USA, AM, bat **$40.00**

Royal 270, vertical, 5$\frac{3}{4}$x3$\frac{1}{2}$x1$\frac{1}{2}$", 1965, plastic, upper right front round dial knob, upper left front round on/off/volume knob, large lower grill area with lower right logo, swing handle, AM, bat **$35.00**

Royal 275, vertical, 1960, plastic, seven transistors, upper right front round dial knob, upper left front on/off/volume knob, large lower lattice grill area with lower right logo, swing handle, AM, bat $35.00

Royal 280, vertical, 5¾x3¾x1¾", plastic, upper right front round dial knob, upper left front on/off/volume knob, lower metal perforated grill area, swing handle, AM, bat $30.00

Royal 285, vertical, plastic, upper right front round dial knob, upper left front on/off/volume knob, lower metal perforated grill area, swing handle, AM, bat $30.00

Royal 300, vertical, 5¾x3⅝x1½", 1958, plastic, upper right front round window dial with right side thumbwheel tuning, left side thumbwheel on/off/volume knob, lower lattice grill area with lower right logo, swing handle, AM, bat ... $35.00

Royal 400, vertical, 5⅞x3⅝x1⅞", 1963, plastic, seven transistors, upper right front round dial knob over large metal perforated grill area, left side thumbwheel on/off/volume knob, swing handle, made in USA, AM, bat $45.00

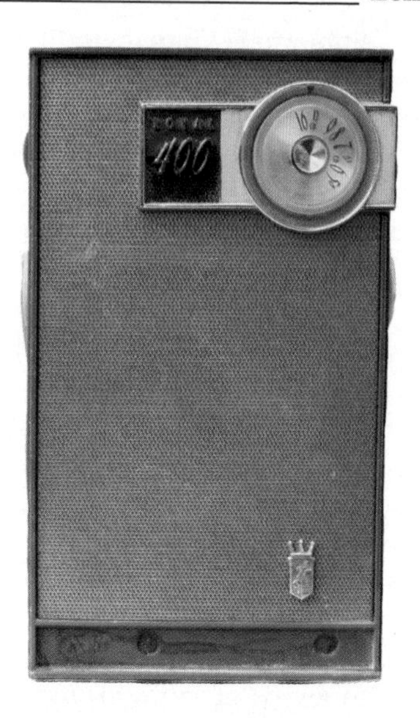

Royal 450, horizontal, plastic, upper left front round dial, lower on/off/volume knob, large right checkered grill area with lower right logo, handle, AM, bat $40.00

Royal 475, horizontal, 1962, seven transistors, upper left front round dial knob, lower left on/off/volume knob, perforated grill area with lower right logo, AM, bat................. $35.00

Royal 490 "Long Distance," horizontal, 1964, leather, seven transistors, upper left front dial knob, lower left on/off/volume knob, right circular grill area with rectangular cut-outs, leather handle, AM, bat **$35.00**

Royal 500

The following six models are only part of the "Royal 500" family. As with other Zenith "Royal" models, the "500" family has areas of confusion in identification because of differences in model numbers, chassis numbers and numbers of transistors – sometimes found in sets that appear to be the same on the outside. These are six of the most popular models with the average collector.

Royal 500 "Hand-Wired," vertical, 5¾x3½x1½", 1955, available in maroon or black plastic, seven transistors, hand-wired chassis (chassis numbers 7XT40, 7XT40Z, 7TX40Z1), the earliest of the Royal 500s, identifiable by its recessed upper front tuning and on/off/volume "owl-eye" knobs – each with a center bar across its diameter – lower round metal perforated grill area, swing handle, AM, bat $150.00

Royal 500, in 1956 three other colors were added to the Royal 500 line – pink, tan, and white – along with the introduction of a printed circuit.

pink **$200.00**
tan **$150.00**
white **$75.00**

Royal 500D, vertical, 5¾x3½x1½", 1959, available in maroon, black, or white plastic, eight transistors, recessed upper front tuning and on/off/volume "owl-eye" knobs with see-through plastic marking protectors, lower round metal perforated grill area, swing handle, AM, bat $75.00

Royal 500E, vertical, 1960/61, available in maroon, black, white, or two-tone plastic, eight transistors, upper right front round dial knob, upper left front round on/off/volume knob, lower round metal perforated grill area, swing handle, AM, bat $60.00

Royal 500H, vertical, 6x3½x1¾", 1962/63, available in black, white, and two-tone gray plastic, eight transistors, step-back top with right thumbwheel dial knob and left thumbwheel on/off/volume knob, large lower oval metal perforated grill area with center logo, swing handle, AM, bat.
black $100.00
white $150.00
two-tone gray $175.00

Royal 500L, vertical, 1964, plastic, upper front horizontal slide rule dial with right side thumbwheel tuning, left side thumbwheel on/off/volume knob, lower metal perforated grill area with lower right logo, swing handle, AM, bat $35.00

Royal 500N, horizontal, 4⅝x5¼x1¾", 1965, plastic, upper front horizontal dial with sliding pointer, right side thumbwheel tuning, left side thumbwheel on/off/volume knob, lower metal grill area with horizontal bars, swing handle, AM, bat **$35.00**

Royal 555 "Sun Charger," horizontal, 4⅝x5¼x1¾", 1966, plastic, up-

per front horizontal dial with sliding pointer, right side thumbwheel tuning, left side thumbwheel on/off/volume knob, lower metal grill area with horizontal bars, swing handle with built-in solar panel, AM, bat $125.00

Royal 650, horizontal, 5⅛x7¼x3", leather, upper left front round dial knob, lower left on/off/volume knob, right leather checkered grill area with lower right logo, leather handle, AM, bat $25.00

Royal 670, horizontal, 5¼x8½x3½", 1964, leather, six transistors, upper right front round dial knob, upper left on/off/volume knob over large grill area with vertical bars, leather handle, AM, bat **$30.00**

Royal 675, horizontal, 5x8½x3⅜", 1960, leather, upper right front round dial knob, large left grill area with vertical bars and upper left on/off/volume knob, handle, AM, bat $30.00

Royal 700, horizontal, 5¼x8⅜x3¼", 1958, leather, seven transistors, upper right front dial knob, upper left front on/off/volume knob, large lower checkered chrome grill area, leather handle, AM, bat $35.00

Royal 710, horizontal, 1960, leather, upper right front dial knob, upper left front on/off/volume knob, large lattice grill area with lower right logo, leather handle, AM, bat **$25.00**

Royal 750, horizontal, 5⅝x8⅞x3¾", 1958, leather, upper right front dial knob, upper left front on/off/volume knob, large lower metal checkered grill area with center logo, leather handle, AM, bat **$30.00**

Royal 755, horizontal/several variations, leather, upper front horizontal slide rule dial, lower metal grill area, two knobs, handle, AM, bat ... $25.00

Royal 760 "Navigator," horizontal, 5½x9x4", 1959, leather, eight transistors, upper right front dial knob, upper left front on/off/volume knob, large metal lattice grill area with lower right and left knobs, top "compass," leather handle, AM, bat $40.00

Royal 780 "Navigator," horizontal, 1960, leather, eight transistors, upper front horizontal two-band slide rule dial, large lower grill area with four knobs and center logo, leather handle, AM, LW, bat **$35.00**

Royal 800, horizontal, 1957, seven transistors, right side dial knob, left side on/off/volume knob, front circular perforated grill area with center logo, pull-up handle, AM, bat **$125.00**

Royal 820, horizontal, 6⅜x11x4½", 1964, leatherette, left front vertical two-band slide rule dial with lower tuning knob, right metal checkered grill area with lower right on/off/volume knob, right side telescoping antenna, pull-up handle, AM, FM, bat $25.00

Royal 950 "Golden Triangle," (bottom right, page 248) vertical/three-sided triangular-shaped clock radio, 9x6", 1960, metal and plastic, seven transistors, first side: round dial with two knobs, second side: round alarm clock face with lower on/off/alarm switch, third side: round perforated metal grill and logo, bottom three-legged base, top handle, clock made in Switzerland, AM, bat $225.00

Royal 1000 "Trans-Oceanic," horizontal, 10⅛x12¾x5", 1958, leatherette/metal/plastic, fold-down front with world map, inner eight band dial and metal perforated grill area, right side band switch, folding handle with built-in telescoping antenna, AM, 7SW, bat $100.00

Royal 1000-1 "Trans-Oceanic," horizontal, 10⅛x12¾x5", 1958, leatherette/metal/plastic, nine transistors, fold-down front with world map, inner multi-band dial and metal perforated grill area, right side band switch, folding handle with built-in telescoping antenna, eight bands, bat **$100.00**

Royal 1000-D "Trans-Oceanic," horizontal, 10⅛x12¾x5", 1958, leatherette/metal/plastic, fold-down front with world map, inner multi-band dial and metal perforated grill area, right side band switch, folding handle with built-in telescoping antenna, nine bands, bat **$100.00**

Royal 2000, horizontal, 10x11⅝x5¼", 1961, 11 transistors, two upper left round dials – one FM, one AM – upper right band and tone knobs, large lower grill area with lower right tuning and on/off/volume knobs, two telescoping antennas, handle, AM, FM, bat **$50.00**

Royal 3000 "Trans-Oceanic," horizontal, 1963, 12 transistors, fold-down front with world map, inner multi-band dial and metal perforated grill area, right side band switch, folding handle with built-in telescoping antenna, AM, FM, LW, 6SW, bat.................. **$145.00**

Royal 3000-1 "Trans-Oceanic," horizontal, 10¼x12½x5⅜", 1963, 12 transistors, fold-down front with world map, inner multi-band dial and metal perforated grill area, right side band switch, folding handle with built-in telescoping antenna, nine bands, bat $145.00

Zephyr

ZR-620, vertical, 4¼x2½x1¼", 1962, plastic, six transistors, upper right front window dial with right side thumbwheel tuning, left side thumbwheel on/off/volume knob, lower metal perforated grill area, made in Japan, AM, bat $40.00

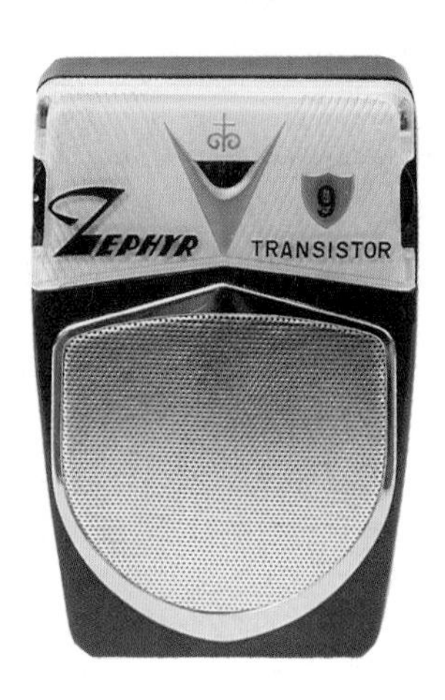

ZR-930, (bottom left) vertical, 4⅝x3 x1¼", plastic, nine transistors, upper front window dial with right side thumbwheel tuning, left side thumbwheel on/off/volume knob, lower metal perforated grill area, rear fold-out stand, AM, bat $45.00

Zohar

MTR-201 "Boy's Radio," vertical, 2½x 4x1", plastic, two transistors, upper right front window dial with right side thumbwheel tuning, left side thumbwheel on/off/volume knob, lower metal perforated grill area, swing handle, AM, bat $30.00

Radio Clubs

The following is a list of antique radio clubs throughout the country. They are always happy to supply potential members with information about their activities and publications. The two clubs listed first are national organizations; the rest are regional.

Antique Wireless Association
P.O. Box E
Breesport, NY 14816.

Antique Radio Club of America
William Dawson
300 Washington Trails
Washington, PA 15301

Alabama Historical Radio Society
4721 Overwood Circle
Birmingham, AL 35222

Antique Radio Club of Illinois
Carolyn Knipfel
RR3, 200 Langham
Morton, IL 61550

Antique Radio Club of Schenectady
Jack Nelson
915 Sherman Street
Schenectady, NY 12303

Antique Radio Collectors Club of Ft. Smith, Arkansas
Wanda Conatser
7917 Hermitage Drive
Ft. Smith, AR 72903

Antique Radio Collectors of Ohio
Karl Koogle
2929 Hazelwood Avenue
Dayton, OH 45419

Arkansas Antique Radio Club
Tom Burgess
P.O. Box 9769
Little Rock, AR 72219

Arizona Antique Radio Club
Treasurer
8311 E. Via de Sereno
Scottsdale, AZ 85258

Belleville Area Antique Radio Club
Karl Stegman
4 Cresthaven Drive
Belleville, IL 62221

Buckeye Antique Radio and Phonograph Club
Steve Dando
4572 Mark Trail
Copley, OH 44321

California Historical Radio Society
Jim McDowell
2265 Panoramic Drive
Concord, CA 94520

North Valley Chapter, CHRS
Norm Braithwaite
P.O. Box 2443
Redding, CA 96099

Carolina Antique Radio Society
Carl Shirley
824 Fairwood Road
Columbia, SC 29209

Central New York/Northern PA Antique Radio Club
Mark Gilbert
711 Elm Street
Groton, NY 13073

Cincinnati Antique Radio Collectors
Tom Ducro
6805 Palmetto
Cincinnati, OH 45227

Colorado Radio Collectors
Bruce Young
4030 Quitman Street
Denver, CO 80212

Connecticut Area Antique Radio Collectors
John Drake
23 E. Wharf Road, Box 942
Madison, CT 06443

Connecticut Vintage Radio Collectors Club
John Ellsworth
665 Arch Street
New Britain, CT 06051

Delaware Valley Historic Radio Club
Radio Attic
P.O. Box 624
Lansdale, PA 19446

Florida Antique Wireless Group
Paul Currie
Box 738
Chuluota, FL 32766

Greater Boston Antique Radio Collectors
Richard Foster
12 Shawmut Avenue
Cochituate, MA 01778

Greater New York Vintage Wireless Association
Bob Scheps
12 Garrity Avenue
Ronkonkoma, NY 11779

Hawaii Chapter/ARCA
ARCA-Hawaii
95-2044 Waikalani Place, C-401
Mililani, HI 96789

Hawaii Historical Radio Club
Kevin Dooley
45 Ala Kimo Drive
Honolulu, HI 96817

Houston Vintage Radio Association
P.O. Box 31276
Houston, TX 77231-1276

Hudson Valley Antique Radio and Phonograph Society
John Gramm
P.O. Box 1, Rt 207
Campbell Hall, NY 10916

Hudson Valley Vintage Radio Club
Al Weiner
14 Prospect Drive
Yonkers, NY 10705

Indiana Historical Radio Society
725 College Way
Carmel, IN 46032

Kentucky Chapter/ARCA
ARCA
1907 Lynn Lea Road
Louisville, KY 40216-2836

Louisiana & Mississippi Gulf Coast Area
F. V. Bernauer
1503 Admiral Nelson Dr.
Slidell, LA 70461

Michigan Antique Radio Club
Jim Clark
P.O. Box 585
Okemos, MI 48864

Mid-America Antique Radio Club
Carleton Gamet
2307 W. 131 Street
Olathe, KS 66061

Mid-Atlantic Antique Radio Club
Joe Koester
249 Spring Gap South
Laurel, MD 20724

Middle-Tennessee Old Radio Club
Grant Manning
Rt 2, Box 127A
Smithville, TN 37166

Mid-South Antique Radio Collectors
Ron Ramirez
811 Maple Street
Providence, KY 42450-1857

Mississippi Historical Radio & Broadcasting Society
Randy Guttery
2412 C Street
Meridian, MS 39301

Nebraska Radio Collectors Antique Radio Club
Steve Morton
905 West First
North Platte, NE 69101

New England Antique Radio Club
P.O. Box 474
Pelham, NH 03076

New Jersey Antique Radio Club
Kathleen Flanagan
92 Joysan Terrace
Freehold, NJ 07728

Niagara Frontier Wireless Association
Gary Parzy
135 Autumnwood
Cheektowaga, NY 14227

Northland Antique Radio Club
P.O. Box 18362
Minneapolis, MN 55418

Northwest Vintage Radio Society
P.O. Box 82379
Portland, OR 97282-0379

Oklahoma Vintage Radio Collectors Club
P.O. Box 72-1197
Oklahoma City, OK 73172

Pittsburgh Antique Radio Society
Richard Harris, Jr.
407 Woodside Road
Pittsburgh, PA 15221

Puget Sound Antique Radio Association
P.O. Box 125
Snohomish, WA 98290-0125

Rhode Island Antique Radio Enthusiasts
Len Arzoomanian
61 Columbus Avenue
North Providence, RI 02911

Sacramento Historical Radio Society
P.O. Box 162612
Sacramento, CA 95816-9998

E. H. Scott Historical Society, Inc.
John Meredith
P.O. Box 1070
Niceville, FL 32588-1070

Society for the Preservation of Antique Radio Knowledge
Harold Parshall
2673 So. Dixie Drive
Dayton, OH 45409

Society of Wireless Pioneers, Inc.
Paul Dane
146 Coleen Street
Livermore, CA 94550

Southeastern Antique Radio Society
David Martin
1502 Wood Thrush Way
Marietta, GA 30062

Southern California Antique Radio Society
C. Alan Smith
6368 Charing Street
San Diego, CA 92117

Southern Vintage Wireless Association
Bill Moore
3049 Box Canyon Road
Huntsville, AL 35803

South Florida Antique Radio Collectors
Victor Marett
3201 N. W. 18 Street
Miami, FL 33125

Vintage Radio & Phonograph Society
Larry Lamia
P.O. Box 165345
Irving, TX 75016

Vintage Radio Unique Society
Jerryl Sears
312 Auburndale Street
Winston-Salem, NC 27104

Western Wisconsin Antique Radio Collectors Club
Dave Wiggert
1611 Redfield Street
La Crosse, WI 54601

West Virginia Chapter, ARCA
Geoff Bourne
405 8th Avenue
St. Albans, WV 25177